BARRON'S
Painless Junior
Science

Wendie Hensley, M.A.
Annette Licata, M.A.

BARRON'S

About the Authors

Wendie Hensley is a Title 1 Resource Teacher for kindergarten through sixth grade in the Baldwin Park Unified School District, Baldwin Park, California. She has an M.A. in curriculum design and instruction and a B.S. in Food and Nutrition. Annette Licata is a third and fourth grade classroom teacher in the same school district. She has an M.A. in reading and a B.S. in Animal Science.

Web Adresses

You should know that addresses on the World Wide Web often change. We've made every attempt to give you the most current addresses. However, by the time you read this book, some of the addresses may no longer work. If you come across such a web address, don't panic. Simply do a key word search to find another site that has the information that you want.

All inquiries should be addressed to:
Barron's Educational Series, Inc.
250 Wireless Boulevard
Hauppauge, New York 11788
www.barronseduc.com

Library of Congress Catalog Card No.: 2007002591
ISBN-13: 978-0-7641-3719-8
ISBN-10: 0-7641-3719-0

Library of Congress Cataloging-in-Publication Data

Hensley, Wendie.
    Painless junior science / Wendie Hensley, Annette Licata.
        p.   cm.
    Includes index.
    ISBN-13: 978-0-7641-3719-8 (alk. paper)
    ISBN-10: 0-7641-3719-0 (alk. paper)
    1. Science—Study and teaching (Elementary)—Juvenile literature. 2. Education, Elementary—Activity programs—Juvenile literature. I. Licata, Annette. II. Title.
    LB1585.H455   2007
    372.3'5—dc22
                                                                        2007002591

PRINTED IN THE UNITED STATES OF AMERICA
9 8 7 6 5 4 3 2

Paper contains a minimum of 15% post-consumer waste (PCW)

Contents

CONTENTS

Icon Key

What You'll Find...
in the chapter.

What You Need to Know
The important terms,
concepts, and facts.

Food for Thought
Interesting information
for you to think about.

Your World and Science
How science relates to you.

Think About It!
A question to test your
understanding.

Let's Try It!
A quiz to check what
you just learned.

Be a Scientist
How you can practice science.

What's Online
Information you can find
on the Internet.

Careful!
Watch out for possible
problems.

v

Introduction

So you want to be an Einstein, a super-genius, a mental marvel? You've come to the right place. This book is divided into three units for the three types of science—life, physical, and earth. Each chapter will introduce a "Big Idea," which is really just a science standard. "What You'll Find" will introduce you to important terms and definitions for that section of the chapter. "Your World and Science" will connect science to your life. "What You Need to Know" is the good stuff. In this section, you will learn important facts about science. "Let's Try It!" is where you have a chance to show what you have learned. There are three levels of questions: Einstein, Super-Genius, and the hardest level, Mental Marvel. The challenging questions in "Think About It" will let you stretch your knowledge limits. And what would a science book be without hands-on activities. "Be a Scientist" lets you have fun with some simple scientific investigations. The chapter ends with "Food for Thought" and "What's Online."

Helpful hints for making the most out of your Painless Junior Science experience:

A highlighter can be a student's best friend. As you read, highlight vocabulary words and important facts.

Post-its are another great tool. You can mark pages that you want to read again or write yourself a question about information you've read.

Keeping a "Remember and Draw" journal can help you understand and remember what you have learned. After you have read the "What You Need to Know" section, you may want to draw simple pictures and write some key words in a notebook or journal to help you remember the new information.

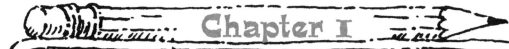

Chapter I

Changes for Survival

THE BIG IDEA

Plants and animals have special body parts and behaviors that help them survive in their habitats.

ANIMAL ADAPTATIONS

What You'll Find

TERMS AND DEFINITIONS

Adaptation

Definition: Adaptations are changes in body parts or behaviors that help animals survive where they live.

In Context: An African elephant's long trunk is an adaptation that helps it to reach tall tree branches for food.

Camouflage

Definition: Camouflage is used to hide something by covering it up or changing the way it looks.

In Context: A chameleon uses camouflage by changing its body color to hide from predators.

Your World and Science

What if you had claws instead of hands? It would make holding your toothbrush very difficult. People have physical adaptations that help them survive every day. Without your thumbs, you could not play a video game or hold your pencil. Try eating a slice of pizza without using your thumbs.

What You Need to Know

Just like you, animals have special body parts that help them survive. These body parts are adaptations. Think about some of your favorite animals. What kind of adaptations do they have? Beaks, claws, whiskers, and stripes are physical adaptations. Here are some examples of how animals use their special body parts:

1. Beaks—Birds use their beaks mainly for eating. An owl will use its sharp beak to tear meat. A parrot will use its strong, curved beak for cracking open nuts.

2. Webbed feet—Webbed feet are perfect for animals that spend time in the water. They help ducks move easily across a pond. Penguins and otters also have webbed feet to help them swim.

3. Claws—A squirrel uses its claws for climbing trees. Crabs have a different kind of claw that they use for protection and holding food. Prairie dogs have a sharp set of claws for digging.

4. Eyes—Placement of an animal's eyes is a special adaptation. Predators, or animals that hunt, such as lions, wolves, and owls have their eyes on the fronts of their faces to help them find their prey. Prey animals, or animals that are hunted, such as rabbits, deer, and zebra have their eyes on the sides of their faces to watch for danger.

5. Camouflage—Camouflage is an adaptation that helps an animal hide from other animals. Walking sticks are insects that look like the tree branches they live on. A tiger's stripes help it to hide in the jungle plants so that it can catch other animals. An octopus can change its skin color to look like the rock it is sitting on. The spotted scorpion fish looks like a rock while it sits on the ocean floor waiting for dinner to swim by.

Let's Try It!

(Answer key page 201)

Einstein Questions

Can you remember what you read? Fill in the blanks. Use your highlighter to find key words in the text above.

1. A crab uses its claws for _____ and for holding food.

2. Parrots use their beaks for _____.

3. Animals with _____ feet move better in the water.

Super-Genius Questions

Match the special feature with its function.

4. An owl's extra large eyes

5. An Arctic fox's white fur

6. A giraffe's very long neck

A. for reaching leaves on tall trees

B. for hiding in snow

C. for seeing at night

Mental Marvel Questions

Choose the correct answer.

7. A tiger's stripes allow it to _____.
   A. look beautiful at a party
   B. hide in jungle plants
   C. scare away enemies
   D. swim on its back

8. A deer has eyes on the side of its head to _____.
   A. see at night
   B. help it run downhill
   C. watch out for danger
   D. be better at playing hide and seek

Think About It

What adaptations does a Pacific walrus have so it can live in and around the ocean? If you need help, go to the websites listed below to find out more.

Be a Scientist

Start your own science journal. Find an animal to **observe** or watch closely. It could be your lazy, old dog or even a bird that has nested in your backyard. What special body parts or adaptations does it have? How does it use these parts? Is it a predator or is it prey? How does it eat and sleep? Draw and label the animal and its special body parts.

Food for Thought

An arrow-poison frog is smaller than your thumb but has enough poison in its skin to kill 1000 people.

Otters live on five continents and are the only ocean mammal to have fur instead of blubber.

What's Online

Try these websites to learn more about animal adaptations.

http://www.ecokids.ca/pub/eco_info/topics/climate/adaptations/index.cfm

http://www.pbs.org/kratts/world/content.html

ANIMALS AND THEIR HABITATS

What You'll Find

TERMS AND DEFINITIONS

Habitat

Definition: A habitat is a place where plants and animals live and grow.

In Context: Pine trees, deer, and raccoons all live together in a forest habitat.

Behavior

Definition: Behavior is the way someone or something acts.

In Context: I was worried about my sister's strange behavior.

Your World and Science

If you were to describe your habitat what would it be like? A room filled with sports equipment, video games, and clothes covering the floor. (Not if your mom can help it.) Your house and neighborhood make up your habitat. It's where you live. Plants and animals also live in habitats.

What You Need to Know

Planet Earth is covered with many types of habitats. Each of those habitats is filled with plants and animals that have special features and behaviors, which help them survive. Here are some examples.

Picture this—You're walking across the hot, dry sand of the desert. There's no water or shade. The sun is beating down on your head. Your throat is so dry that you're dreaming of swimming pools and ice-cold lemonade. Your sweat-soaked shirt is sticking to your back. The sand is so hot you can feel it through the bottoms of your shoes. How do you survive here? You'd probably hike back to the motor home and pull a soda out of the ice chest. But if you were a jackrabbit, it wouldn't be so easy.

You might think that nothing could survive here, but many plants and animals call this place home. How do they do it? They all have special features and behaviors that help them live in this desert habitat.

An average desert gets less than 14 inches of rain per year. The desert has very hot days and very cold nights. It is covered with sand and low-growing plants.

• Desert Tortoise

Special feature—Tortoises have large, thick legs with strong claws that help them dig.

Special behavior—Tortoises burrow into large holes to keep out of the sun and the cold night air.

- Gila Woodpecker

  Special feature—Woodpeckers have sharp beaks to help them drill holes in a cactus and for catching and eating insects.

  Special behavior—Woodpeckers nest high inside a cool, safe cactus.

- Barrel Cactus

  Special feature—The cactus is covered with sharp spines that protect it from being eaten.

  Special behavior—The cactus likes eating cold pizza for breakfast. Just kidding. Wanted to make sure you were paying attention.

Picture this—You're walking through a dark, thick forest where the air is wet and sticky. Your hair and clothes are stuck to your skin. The thick canopy of leaves above you blocks the sunlight. An ant walks over the toe of your shoe, and it is nearly as big your grandma's toy poodle. You have entered the dark green habitat known as the rain forest. And the animals and plants that live in this habitat are amazing.

The average rain forest is hot and wet all year long. It has very tall trees that stay green all year. Many plants grow on the forest floor.

- Orangutan

  Special feature—Orangutans have long arms for swinging and climbing.

  Special behavior—Orangutans build their nests up high in trees for safety.

- Tree Frogs

  Special feature—Tree frogs have toes with round pads that make a sticky goo, which helps them climb.

  Special behavior—Tree frogs are nocturnal, which means they are up at night. This helps them hide from predators.

- Lianas

  Special features—Lianas are vines with extra long roots that help them to climb trees.

  Special behavior—Lianas climb tall trees to reach up through the heavy canopy to find sunlight.

Picture this—You're covered from head to toe in warm winter clothes as you walk slowly across slippery ice. Your nose is so frozen that if you sneeze it might break off. You see nothing but an endless sea of white snow and ice. There are no trees and very few living things. You wonder if anything lives here on the tundra.

An average tundra is flat with no trees. There is a frozen layer under the ground that never melts.

- Arctic fox

  Special features—Arctic foxes are covered in thick, white fur even on their ears and the bottoms of their feet. The fur keeps them warm and makes them hard to see in the snow.

  Special behavior—Arctic foxes travel longer distances than any other land animal except humans.

- Puffin

  Special feature—The puffins' thick, waterproof feathers protect them and keep them warm while diving in the ocean.

  Special behavior—Puffins spend a lot of time floating on the ocean surface and cleaning their feathers.

- Arctic poppy

  Special feature—The Arctic poppy's cup-shaped flower helps it catch sunlight.

  Special behavior—The poppy flower turns to follow the sun and attract insects to its warm petals.

Picture this—You're swimming in the salty water of the ocean. You must keep moving your arms and legs so the waves do not drag you away. Behind every rock, a deadly predator might be hiding ready to make a meal of you. And the deeper you swim, the darker and colder it gets.

There are no average oceans. The temperature, sunlight, and salt are different in each ocean.

- Shark

Special feature—Sharks have light, soft bones to keep them floating.

Special behavior—Many sharks hunt by staying very quiet until a fish swims past.

- Clownfish

  Special feature—Clownfish have bright colors, which attract other fish to swim closer.

  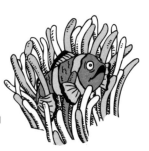

  Special behavior—The clownfish make their home in or near the stinging anemone so that predators that swim near are stung.

- Seaweed

  Special feature—Some types of seaweed have large leaves for catching sunlight.

  Special behavior—Some types of seaweed whip other types of seaweed out of their space.

Let's Try It!

(Answer key page 201)

Einstein Questions

Match the following animals to their habitats.

1. <u>Puffins</u> have waterproof feathers to keep them warm.

   A. desert

2. <u>Clownfish</u> live inside the sea anemone.

   B. tundra

3. The <u>tortoise</u> uses its claws to dig large holes to keep out of the sun.

   C. forest

4. <u>The orangutan</u> has long arms for swinging and climbing.

   D. ocean

Super-Genius Questions

Choose the correct answer.

5. The North American porcupine's body is covered with sharp quills. What do you think they use these sharp quills for?
   A. for shade to cool the animal
   B. to pop small children's balloons
   C. for protection from predators
   D. to catch food

6. An anteater has a very long, sticky tongue? What does it use this tongue for?
   A. to clean its fur
   B. to catch ants from their hole
   C. to scare predators away
   D. to lick a Popsicle in summer

Mental Marvel Questions

Choose whether the animal characteristic below is an adaptation that is a physical feature (PF) or special behavior (SB).

7. A thick layer of waterproof fur   PF or SB
8. A long bristly tongue   PF or SB
9. Building a nest of dirt and moss   PF or SB
10. Furry feet bottoms   PF or SB
11. Floating silently in water   PF or SB

Think About It

Why would a fox living in the forest have long claws for digging? If you need help go to the websites listed below to find out more.

Be a Scientist

Observe a habitat and record or write down what lives there in your science journal. Go out to your backyard or the park. What kinds of plants live there? List them in your journal. What kinds of animals live there? Don't forget to include insects. You will also want to include any nonliving things in the habitat such as rocks. Draw a picture of your habitat including all the things you have observed.

Food for Thought

You're back in your own habitat, namely your bedroom. How would an animal adapt to your habitat? It might need extra long legs to step over the piles of dirty clothes on your floor. It would need thumbs so that it could play your video games. It might have an extra good sense of smell to find that grilled cheese sandwich you've been missing since last week. Of course, it would be cool if the animal had a special behavior that made it clean up clothes and toys. In other words, this animal would be a lot like you—except maybe the cleaning up part.

After the Japanese sea squirt finds a cozy rock to live on in its ocean habitat, it eats its own brain!

The Couch's spadefoot toad lives in the desert and comes out of the ground in rainy July. It stays underground the other 11 months of the year.

What's Online

Try these websites to find out more about animals and their habitats.

http://www.hitchams.suffolk.sch.uk/habitats/index.htm

http://www.nationalgeographic.com/geographyaction/ habitats/deserts_tundra.html

EVOLUTION AND EXTINCTION

What You'll Find

TERMS AND DEFINITIONS

Environment

Definition: Your **environment** is your surroundings or everything around you that is part of your life.

In Context: Penguins live in an icy, wet **environment**.

Climate

Definition: **Climate** is the usual weather conditions of a place.

In Context: Deserts have a very dry **climate**.

Extinct

Definition: **Extinct** means one kind of animal has died out completely.

In Context: Dinosaurs are **extinct** or no longer alive.

Evolution

Definition: **Evolution** is constant change over a long period of time.

In Context: **Evolution** has changed plants and animals over millions of years.

Your World and Science

What if over millions of years our world changed? What if most of our land was covered with water? This slow rise of the sea would take hundreds of years. How would people in the future adapt to their new world? Would feet with five separate toes be as useful in this new world of water as they were on land? When you are in a pool you can use a snorkel to breathe underwater. Is it possible our lungs would adapt so that people could breathe underwater like fish? And if the rise of the oceans was more sudden, and people did not have time to adapt, would they survive at all? These are all scientific questions to wonder about.

What You Need to Know

EVOLUTION

Some types of animals change a lot over a long period of time. A perfect example of this is the horse. The first horse walked the Earth about 55 million years ago. It did not look anything like the horses we see today.

Millions of years ago, the Earth was covered with jungles filled with tall, leafy plants. The horse was a small fox-like animal that ate berries. It had teeth that allowed it to eat different types of food. It was short enough to hide in the thick jungle plants.

Over millions of years, the climate changed and became drier. The jungle was replaced by more grasslands. The horse's teeth changed to allow it to eat grass. Because there were no more jungles to hide in, the horse's legs grew longer so it could run faster.

All these changes took place over 55 million years. However, this is not the end of the story. The climate, Earth, and animals continue to change. Maybe in the future, horses will have wings!

"TA DA!"

Some types of animals such as the crocodile and shark have not changed in millions of years. These animals have not needed to change.

"Hey Fred, haven't seen you in years. You haven't changed a bit."

EXTINCTION

Our planet is billions of years old. Many changes have taken place during that time. Climate changes can greatly affect life on Earth. Dinosaurs once walked our great planet. But now they are extinct. No one knows exactly why the dinosaurs died off, but many scientists have tried to explain their disappearance. Did a giant meteor crash into Earth, blocking out the sunlight and causing the plants and animals to die? Or did an ice age kill some plants and animals? Is it possible that a terrible sickness caused these giant animals to die? These are only three of the questions being asked about dinosaur extinction.

Let's Try It!

(Answer key page 201)

Einstein Questions

Can you remember what you read? Use your highlighter to find key words.

1. Over time, the early horse's legs grew longer so it could _____.

2. _____ and _____ are two animals that have not changed over millions of years.

3. Scientists are unsure how the dinosaurs became _____.

Super-Genius Questions

Decide if the statement is **true** or **false**.

4. All animals change over a long period of time. True or False

5. The teeth of horses changed as plants changed. True or False

6. Scientists are sure that an ice age killed all the dinosaurs. True or False

Mental Marvel Questions

Choose the correct answer.

7. Cave fish live in caves with no light. Over time, a layer of skin has grown over their eyes causing them to become blind. Why do you think this has happened?
   A. They like to sleep all day.
   B. They don't like to look at each other.
   C. They do not need to use their eyes.
   D. All of the above

8. What can people do to stop the extinction of animals?
   A. Take better care of our planet.
   B. Take care not to destroy animal habitats.
   C. Recycle bottles and cans.
   D. All of the above

Think About It

In "What You Need To Know" you learned about the evolution of the horse and the changes in its teeth and legs. These changes helped the horse survive in its new environment. What other physical changes to the horse helped it to survive? If you need help go to the websites listed below to find out more.

Be a Scientist

Getting back to our future world filled with water, what would land animals do in this new world? In your science journal pick one animal, such as a dog, cat, horse, or moose. How would the animal you picked change or evolve to live in this new habitat of water? Draw a picture of your new improved animal. Don't forget to include things like how it would breathe and move and what it would eat.

Food for Thought

Up to 50–250 animal species become extinct each day. In the U.S. alone, 1263 species are nearly extinct.

What's Online

Try these websites to find out more about evolution and extinction.

http://www.chem.tufts.edu/science/evolution/HorseEvolution.htm

http://www.bbc.co.uk/sn/prehistoric_life/

Chapter 2

The Circle of Life

THE BIG IDEA

All living things need energy to live and grow.

PLANTS AS AN ENERGY SOURCE

What You'll Find

TERMS AND DEFINITIONS

Photosynthesis

Definition: **Photosynthesis** is the process where plants use sunlight to turn water and carbon dioxide from the air into sugar and starches, which are food.

In Context: If a green plant does not have sunlight and water for **photosynthesis,** the plant will die.

Chlorophyll

Definition: **Chlorophyll** gives plants their green color and makes photosynthesis possible.

In Context: Without **chlorophyll,** plants would not be able to make their own food from water, carbon dioxide, and sunlight.

Your World and Science

What would happen if all the grocery stores and restaurants closed, and your kitchen was empty of food? You would become very hungry and lose energy. Remember in Chapter 1 when we discussed the extinction of dinosaurs? Some scientists believe that sunlight was blocked when a giant meteor crashed into the Earth making a huge cloud of dust. Although the plants and dinosaurs probably did not die immediately, they would eventually have starved to death because plants could not grow without sunlight. This section will explain how sunlight and plants help put orange juice in your glass, tacos on your plate, and oxygen in your air through a process known as photosynthesis.

What You Need to Know

Picture this—You wake up one morning and discover you are now a pumpkin plant. At first you freak out, but then you stretch your leaves out in the warm sun and decide this could be an easy life. Not so fast. You are a green plant now, and there is lots of work to do. You've got pumpkins to grow. You wiggle your toes, only to discover that now they are roots. Your roots search under the soil for some water to drink. The water climbs all the way up to your leaves. In the meantime, your leaves are breathing in carbon dioxide from the air. Now that you have water and carbon dioxide, the chlorophyll in your leaves can get busy working with the sunlight to make sugars. You're not done yet. This process of making sugars goes on and on until you grow nice, strong vines. Now you have enough sugars that you can begin to store the extra sugars in the form of starches. It's time to start growing round, orange pumpkins.

Let's break it down with the science terms. For photosynthesis to happen:

Plants have:

- **Chlorophyll** is the green color that works with the sun's energy to make food in the form of sugar for the plant to grow.

- **Roots** take in water for the plant to use in photosynthesis.

Plants need:

- **Sunlight** works with chlorophyll already in the plant to help it make food in the form of sugars and starches.

- **Water** from rain or the garden hose is taken into a plant through its roots. It combines with carbon dioxide during photosynthesis to make sugars and starches.

- **Carbon dioxide** is taken in from the air around the plant through tiny holes in the leaves. It combines with water during photosynthesis to make sugars and starches.

Plants make:

- **Sugars** are made from carbon dioxide, water, and sunlight during photosynthesis.

- **Oxygen** is made during photosynthesis and released into the air we breathe.

You might ask: What does all of this have to do with me? Luckily for humans and animals, plants make far more food energy than they need during photosynthesis. The extra sugars and starches are stored in fruits, nuts, vegetables, roots, seeds, and leaves. If it weren't for sunlight and green plants, you would never be able to drink a glass of orange juice. Cows would not have any grass to eat if photosynthesis didn't happen. This would mean no milk or hamburgers either.

Let's Try It!

(Answer key page 202)

Einstein Questions

Can you remember what you read? Fill in the blanks. Use your highlighter to find key words in the text above.

1. _____ gives plant leaves their green color.

2. Sunlight works with chlorophyll to help plants make food in the form of _____.

3. Tiny _____ in the plant's leaves help it to take in carbon dioxide from the air.

Super-Genius Questions

Match the word with its definition.

4. Chlorophyll

A. process where plants make their own sugars and starches

5. Photosynthesis

B. chemical produced during photosynthesis and released into the air

6. Oxygen

C. plant parts that take in water for the plant to use for photosynthesis

7. Roots

D. the green color in plants

Mental Marvel Questions

Choose the correct answer.

8. During photosynthesis, a plant makes sugar and starches from _____.
   A. soil and water
   B. water and carbon dioxide
   C. brownies and cookies
   D. roots and leaves

9. During photosynthesis, light from the sun works with the green color in a plant known as _____.
   A. carbon dioxide
   B. oxygen
   C. candy
   D. chlorophyll

Think About It

Are mushrooms considered part of the plant kingdom? Why or why not? If you need help, go to the websites listed below to find out more.

Be a Scientist

Ask your parents if you can buy two small sun-loving green plants from the store. Find two small plants that are the same type and the same size. **Label** one plant "Sun" and the other plant "No Sun." **Measure** each with a ruler to see how tall they are, and record the measurements in your journal. You will also want to draw a picture of each plant. Remember to color the picture because the color of the leaves might change by the end of the experiment. Give each plant equal amounts of water each day. This way only one thing (or **variable**) will be different in your experiment. Do not use too much water though. If the soil becomes soaked, the plants may die. Place the "Sun" plant in direct sunlight. Place the one labeled "No Sun" in a dark place, like your closet. **Measure** the plants every two days, and **record** the amount of growth each plant makes. You might also want to draw more pictures after a week or so. Write about your **observations**. Are both plants growing at the same rate? Are they both still green? Your number or **data chart** might look like this:

Plant	Day 1	Day 3	Day 5	Day 7	Day 9	Day 11
SUN	3 inches					
NO SUN	3 inches					

Food for Thought

Did you ever wonder where leaves get their beautiful orange and yellow colors in fall? In the fall, plants begin to get ready for winter. There is not enough sunlight in the cold months for photosynthesis to happen, so most plants stop making sugars and starches. Green chlorophyll begins to disappear from the leaves as the weather gets colder. Small amounts of yellow and orange are already in the leaves. Once the chlorophyll is gone, the yellow and orange colors show through the leaves.

Since plants in the desert get plenty of sunlight but very little water, they must find other ways to store water. Some desert plants take water in with their thick leaves and needles. Some drop all of their leaves during very dry periods. And some desert plants only last for a short season in the fall after the summer heat cools down.

What's Online

Try these websites to find out more about plants as an energy source.

http://www.ftexploring.com/photosyn/photosynth.html

http://www.factmonster.com/ce6/sci/A0860378.html

FOOD CHAINS AND FOOD WEBS

What You'll Find

TERMS AND DEFINITIONS

Producer

Definition: Producers are plants that make their own food in the form of sugars and starches during photosynthesis.

In Context: Producers or plants use sunlight to make their own food.

Consumer

Definition: Consumers are animals that must eat plants and/or other animals to survive.

In Context: Consumers or animals get their food energy from eating plants and other animals.

Herbivore

Definition: Herbivores are animals that eat only plants.

In Context: An herbivore does not eat other animals.

Carnivore

Definition: **Carnivores** are animals that mostly eat meat or other animals.

In Context: A cat is a **carnivore** because it would love to eat your goldfish.

Omnivore

Definition: **Omnivores** are animals that eat both plants and animals.

In Context: Only an **omnivore** would enjoy eating a bacon, lettuce, and tomato sandwich.

Your World and Science

You wake up this morning excited to go skateboarding with your friends, but before you can make it out the door, you fall on to the nearest chair, too tired to play. What happened? You have no energy. What? You skipped breakfast? Where do you think you get energy? That's right. Food!

Without food energy, you cannot grow, learn, or play. At school, you might have noticed that's its harder to pay attention right before lunch. It's not because math is boring or because the kid next to you is throwing erasers across the room. It's because you need more energy. You need food. Without food energy, you can't focus on your schoolwork or play at recess.

What You Need to Know

Like you, all living things need energy to live and grow, and that means food. But not all living things get their energy in the same foods. A food chain shows us where each living thing gets its food energy.

PRODUCERS

All energy comes from the sun, but animals cannot use the sun's energy without the help of plants. Plants are the start of any food chain. As you learned in the first section of this chapter, plants can use energy from the sun to make their own food through photosynthesis. Plants are producers because they make their own food.

CONSUMERS

If plants are the producers in a food chain, then animals are the consumers. Consumers are animals that eat or consume plants or other animals to get their energy. There are three types of consumers.

Herbivores are animals that eat only plants. Animals like the horse, giraffe, deer, and elephant are in this group. These animals eat grasses, leaves, and even tree bark, but they never eat other animals.

Carnivores are animals that mostly eat other animals. In Chapter 1, you learned that animals that hunt other animals for food are called predators. Large predators like lions, tigers, and sharks are all carnivores because they eat other animals. Smaller animals like hawks, cats, and otters are also predators and carnivores.

Omnivores are animals that eat both plants and animals. You belong to this group because you eat both the broccoli your mom makes you for dinner and the hamburger your dad buys you for lunch. A grizzly bear is another example of an omnivore. In their forest habitat, bears might eat berries off a bush in the morning and fish from a cool stream in the afternoon.

Food Chains

Picture this—You are on a photo safari in Africa. Through the lens of your camera, you are observing a herd of wildebeest and zebra grazing peacefully on the tall grass of the African plain. Suddenly their heads pop up. What is it? What do they see? Now you see it too. A lion crouches at the edge of the field, waiting to make its move. As you watch, the lion leaps into action and chases its prey. The huge herd thunders across the plain. A slower, weaker zebra falls behind the herd and into the claws of the hungry lion. The zebra will make an excellent meal for the lion and her cubs. And while it's not pretty, you have just observed a food chain in action.

Plants or producers are the first link in all food chains. The tall grasses on the African plain are the producers in this story.

The first consumers in a food chain would have to be an herbivore or an omnivore. In the story, the zebras and wildebeest are the herbivores in the food chain.

The next link in the food chain would be an animal or consumer that mostly eats other animals. A large carnivore is usually at the top of the food chain, and, in this case, that would be the lion.

Careful!

Food chains come in different lengths. Although the average food chain has three to four links, some may be shorter, and some may be longer.

Now let's look at food chains from another point of view.

Picture this—You are a mesquite bush sitting all alone in the hot desert sun. You use your very long roots to find underground water to help grow your seedpods. Suddenly a gust of wind drops several of your seedpods to the ground. Then your pods are covered with tiny harvester ants. You have just become part of a desert food chain.

Now you are a harvester ant. You are busy helping your 10,000 brothers and sisters carry seedpods back to the anthill. But beware! You have entered a desert food chain, and you are the next link! There you are minding your own business, wiggling your reddish brown body in the hot desert sun as you carry a heavy seedpod. A large, ugly shadow looms over you. Next thing you know, you are gone.

Now you are a horned lizard. You are strutting proudly in the hot sun thinking it is great to be a horned lizard. After all, you look like a mini dragon, and that is pretty cool. The sand you are crawling through is extra hot today, but that doesn't bother you. You are hungry. Up ahead you can see a soft mound of sand, which means dinner is only a few feet away. And bingo! Harvester ants! They are too busy with their work to notice you. Yum! With your tummy full, you decide to take a nap in the hot sun. You have no worries in the world. Or at least that is what you think!

You are now an angry hawk circling the dry desert below. You had a plump rat in your claws, but he got away. Now you're really hungry. You are high in the sky drifting on the warm breeze when you spot something

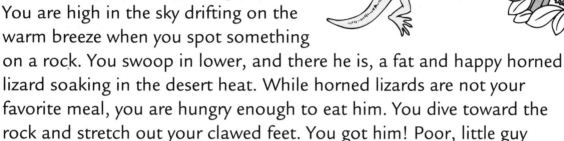

on a rock. You swoop in lower, and there he is, a fat and happy horned lizard soaking in the desert heat. While horned lizards are not your favorite meal, you are hungry enough to eat him. You dive toward the rock and stretch out your clawed feet. You got him! Poor, little guy never saw it coming.

FOOD WEBS

A food chain is a pathway that shows where living things get their food. A food web connects different food chains. The grass in Africa that the zebra ate also feeds wildebeest, antelope, and elephants. In turn, lions, cheetahs, and hyenas eat these animals. Most food chains only have three to four links or living things on them, but food webs will include all the animals living in a habitat. An example of a food web would be:

Let's Try It!

(Answer key page 202)

Einstein Questions

Can you remember what you read? Fill in the blanks. Use your highlighter to find key words in the text above.

1. _____ always start a food chain.

2. There are usually no more than_____ to _____ links in a food chain.

3. Consumers are animals that eat _____ and/or _____.

Super-Genius Questions

Choose the correct answer.

4. Why do all food chains start with plants?
   A. Green is good for you.
   B. Plants can change sun energy to food energy.
   C. Plants taste better than animals.
   D. There are more plants than animals.

5. A food web is _____.
   A. a really big home for a spider
   B. a place where plants get food
   C. different food chains connected together
   D. a web that has a moth caught in it

Mental Marvel Questions

Arrange these plants and animals in a food chain.

6. lizard   cactus   hawk

     _____ ⟶ _____ ⟶ _____

7. insect   flower   bear   salmon

     _____ ⟶ _____ ⟶ _____ ⟶ _____

8. killer whale   plankton   seal   fish

     _____ ⟶ _____ ⟶ _____ ⟶ _____

Think About It

What are some top-level carnivores in an ocean food web? If you need help go to the websites listed below to find out more.

Be a Scientist

Remember back to Chapter 1, you **observed** a habitat and **recorded** what lived there in your science journal. Go back to your journal and look at the plants and animals you **recorded**. Do you see a food chain in your habitat? Can you **identify** or find a food web? Draw them in your journal. Start with the plants and finish with the largest carnivore you can find in your habitat. It might be a hawk, raccoon, or even your family dog or cat.

Food for Thought

You are back at your house and you still want to go skateboarding, so you need energy. You might eat a nice breakfast of ham and eggs or maybe a bowl of cereal. All of these things come from plants and animals. They will give you the energy you need to play and grow strong bones and muscles. Remember you are the top of the food chain because nothing eats people as a regular part of its diet.

Elephants need 220 to 440 pounds of food every day.

A blue whale, the largest mammal on the planet, will eat up to 8000 pounds of krill every day for 4 months. The rest of the year is spent traveling and breeding.

What's Online

Try these websites to find out more about food chains and food webs.

http://www.gould.edu.au/foodwebs/kids_web.htm

http://www.vtaide.com/png/foodchains.htm

DECOMPOSERS

What You'll Find

TERMS AND DEFINITIONS

Decomposer

Definition: **Decomposers** are living things that feed on and break down dead plants or animals.

In Context: When a leaf dies, a **decomposer** breaks it down into the soil.

Fungi

Definition: **Fungi** are living things that feed on dead plants and do not have chlorophyll to make them green.

In Context: **Fungi** help green plants by breaking down dead material in the soil.

Bacteria

Definition: **Bacteria** are living things that can only be seen through a microscope and are found in soil, water, and the bodies of plants and animals.

In Context: Scientists study good and bad **bacteria** through a microscope.

Recycle

Definition: **Recycle** means to reuse resources.

In Context: I **recycle** aluminum cans so they can be made into new aluminum cans.

Nutrient

Definition: **Nutrients** are sources of food energy for plants and animals.

In Context: A plant can get **nutrients** it needs from the soil around it.

Your World and Science

When you're done with a can of soda or bottle of water, what happens to it? Do you throw it into the trash, or do you recycle it? Recycling is a great way to reduce the amount of trash we collect. We use less of the Earth's resources and keep our planet cleaner when we recycle. Recycled things get broken down into tiny pieces and reused to make new things. If we didn't recycle, in time, there would not be any material left to make cans or bottles. And there wouldn't be any place left to put all the trash.

The same is true in the world of plants and animals. They have been recycling since the world began. Decomposers are the recyclers in nature. We would be up to our knees in dead plants and animals without the help of decomposers.

What You Need to Know

We have talked about producers and consumers as part of the food chain. But there is one more link in the food chain that we have not talked about. Decomposers are the last link in the food chain. They break down dead plants and animals into nutrients that can then be used by plants. As decomposers eat, they break down dead material into nutrients. These nutrients then go back into the soil, air, and water to be reused by plants. There are three types of decomposers.

Fungi are mushrooms and molds. They are not plants because they do not have chlorophyll, and they do not make their own food. They get their food by taking in the nutrients from the dead plants and animals they break down. The green fuzzy spots you see on bread and cheese when it is old is a fungus. This green fungus has already started breaking down your bread and cheese. The largest living thing on the planet is a fungus called the honey mushroom. This giant fungus is found in Oregon and is more than 3 miles wide.

Bacteria are some of the smallest life forms on the planet. There are more bacteria in a single handful of dirt than there are people on our planet. If you leave uncooked meat on the kitchen counter overnight, it will begin to smell funny. (Unless your dog grabbed it while you weren't looking.) It smells funny because bacteria have begun to break down the meat. Have you ever taken a cucumber out of your refrigerator only to find that it feels slimy? Those are bacteria at work decomposing the vegetable. This also happens in nature. Not just in your kitchen.

Insects can act as decomposers in nature. When an animal dies, insects are the first decomposers to get to work. Flies, cockroaches, and ants are examples of insects that help break down plants and animals. If you leave your hamburger on a picnic table what happens to it? Soon it will be covered with flies. They have landed both to have a snack and to begin their work as decomposers.

Let's Try It!

(Answer key page 202)

Einstein Questions

Can you remember what you read? Use your highlighter to find key words.

1. Fungi, bacteria, and insects are all _____.

2. Fungi are _____ and _____.

3. Bacteria are some of the _____ life forms on the planet.

Super-Genius Questions

Decide if the statement is **true** or **false**.

4. If you leave meat out of the refrigerator for too long, it starts to smell because it has been overcooked. True or False

5. Decomposers are the first link in the food chain. True or False

6. When an animal dies, insects are the first decomposers to get to work. True or False

Mental Marvel Questions

Choose the correct answer.

7. When you recycle you _____.
   A. return home on your bicycle
   B. help to reduce the amount of trash on the planet
   C. help flies to feed their families
   D. burn trash in a campfire

8. The three types of decomposers are _____.
   A. plants, flies, and mold
   B. hamburger, pickles, and mustard
   C. mold, fungus, and sunlight
   D. fungi, bacteria, and insects

Think About It

Are earthworms decomposers? If you need help, go to the websites listed below to find out more.

Be a Scientist

With the help of your parents, you can watch decomposers in action by building your own compost pile. Compost is broken-down plant material used to help gardens grow. A compost pile begins with a mixture of plants, soil, and water. To start a compost pile, you need a container like a trash can without a lid placed in a safe area in your yard. Start by putting a layer of dead leaves and twigs on the bottom of your container. Then add a layer of fresh grass and old fruits and vegetables. Make sure to sprinkle water on each layer. Each week add a new layer of brown material and green material. Don't forget to water it. After two weeks you can add in a bucket of soil. This soil will have the decomposers needed to break down the green and brown plant layers into rich compost for your garden. In your journal **record** what you used in your compost pile and how it looked each week.

Food for Thought

If decomposers stopped working for a month, the Earth would be covered with a layer of flies 20 feet deep.

Humans can have up to one million bacteria in their bodies at one time.

What's Online

Try these websites to find out more about decomposers.

http://www.nhptv.org/natureworks/nwep11.htm

http://www.qrg.northwestern.edu/projects/marssim/simhtml/info/whats-a-decomposer.html

Chapter 3

Depending on Each Other

THE BIG IDEA

All living things depend on each other and their environment to live and grow.

ECOSYSTEMS

What You'll Find

TERMS AND DEFINITIONS

Ecosystem

Definition: An ecosystem is a system of plants and animals living together and depending on each other and their non-living environment.

In Context: Frogs, insects, and plants all depend on each other in a pond ecosystem.

Physical environment

Definition: Physical environment includes everything in an area that is non-living such as water, soil, and minerals.

In Context: A rattlesnake likes to live in a physical environment that is hot and dry.

Minerals

Definition: Minerals are nutrients that are not animal or plant.

In Context: The salt that you put on your fries is a mineral.

Your World and Science

Have you ever taken a hike in the mountains or a walk on the beach? Did you know that when you climb a trail or walk around a tide pool at the beach, you are walking through many ecosystems? Even the field in your schoolyard has ecosystems. Ecosystems can be as small as a puddle of rain in the middle of the soccer field or as big as the entire planet. And like the puddle in the soccer field and planet Earth, ecosystems are always changing. In Chapter 1, you learned that plants and animals can live together in the same habitat. Now you will learn how plants and animals form an ecosystem by depending on their habitat and on each other.

Careful!

The habitat is the place where plants and animals live. The ecosystem is the plants and animals depending on their habitat and each other to live.

What You Need to Know

In the definition above, we learned that an ecosystem contains plants and animals living together and depending on each other and their environment. An ecosystem can be very small and simple with only one or two forms of life. The puddle in the middle of the soccer field is a good example of this. If you look carefully at a puddle, you may observe tiny insects moving in the water. They are larva (babies) from insects that wait for rain as a chance to lay their eggs. These insects depend on the water, sun, and air to make new life. The puddle is also a good example of an ecosystem that is always changing. As the sun dries the puddle, the tiny ecosystem disappears.

An ecosystem can be very large and contain a number of different plants and animals. It can even include a number of smaller ecosystems. A forest is a good example of a habitat and large ecosystem that contains many smaller ecosystems. In the forest, a mountain stream is an ecosystem that has living and non-living parts that depend on each other. In the stream, fish, frogs, and insects depend on the plants, water, and rocks for food and shelter. The plants and animals that live in and around a mountain stream depend on the water and each other to survive. Deeper in the forest, a large fallen tree would be home to a completely different ecosystem. Insects may live beneath the fallen tree trunk eating dead plant material. Plants and fungi that like shade may begin to grow in the shadow of the big fallen tree. A small animal like a squirrel may decide the fallen tree is the perfect place to build a home.

Even though there are many kinds of ecosystems, they all have two parts: living parts and non-living parts.

Non-living parts:

- Soil type in an area determines the type of plants that will grow in the ecosystem.

- Water in an ecosystem can be in the form of a lake or stream. Or it can be moisture in soil from rain, snow, or fog.

- **Nutrients** that are non-living are called minerals. Plants and animals need minerals to survive. They get minerals from water, soil, and the food they eat.

- **Energy** from the sun is the first source of energy in food chains. Ecosystems contain food chains that begin with energy from the sun. Plants then use this energy for photosynthesis.

Living parts:

- **Plants, fungi, and mold** are part of a healthy ecosystem. They provide a food source for consumers or animals and insects. The types of plants that grow in an ecosystem depend on the non-living parts listed above.

- **Insects** are important as food sources and as decomposers in an ecosystem. Some insects, like bees, are necessary for plant growth.

- **Animals** are part of an ecosystem. The type of animals living in an ecosystem depends on the surrounding habitat, plants, insects, and all of the non-living parts listed above.

An ecosystem is always changing. Sometimes a change can be bad for the ecosystem. If a plant or animal enters an ecosystem in which it does not belong, the whole ecosystem may suffer. If someone decided to let loose their pet rabbits into a park, the rabbits may eat plants that were supposed to be for animals that live in the park's ecosystem. Then the park animals may not have enough food to eat. Plants or weeds that are not supposed to be a natural part of an ecosystem can cause other plants to die, changing the source of energy in an ecosystem. Weather that is too dry or too wet can also hurt the health of an ecosystem.

Examples of Ecosystems

Picture this—Get your snorkel and fins, you're going diving! One of Earth's most exciting and colorful ecosystems is the coral reef. This ocean ecosystem is home to many forms of life. As you make your way through the clear, blue water, you look down at a forest of colorful, hard coral. Coral is living, and it depends on algae and plankton for energy. Floating in and around the sharp edges of coral are the many sea animals that call it home. Through your mask, you might see crabs, rays, lobsters, and turtles. As you swim over the reef, you see more plant and animal life. Clusters of seagrass, an ocean plant, grow behind the protective wall of jagged coral. There you'll see turtles, fish, and sea cucumbers using the seagrass for food and shelter. In one thick clump of seagrass, you spot a lobster laying its eggs. Careful, you don't want to run into the thick roots of the mangrove trees that grow near the coral. Coral reefs provide shelter for the mangrove trees. The

mangrove trees help the coral by keeping the ocean floor in place. The mangroves also provide shelter for marine wildlife including birds. As you see, the coral reef is a busy and important ecosystem. Many people who live near the coast depend on coral reefs for food and to keep the ocean from washing away the shore. Unfortunately, coral reefs are ecosystems that are in danger of disappearing. Rising ocean temperatures, pollution, and damage by boats are all problems for the coral reef.

Let's Try It!

(Answer key page 203)

Einstein Questions

Can you remember what you just read? Fill in the blanks.
Use your highlighter to find key words in the text above.

1. An ecosystem can be very large and contain a number
   of _____ and _____.

2. Insects are important as food sources and as
   _____ in an ecosystem.

3. Rising ocean temperatures, _____,
   and damage by boats are all problems for the coral
   reef.

Super-Genius Questions

Circle the thing that would **not** belong in the ecosystem.

4. Pond ecosystem—mosquitoes, frogs, lily pad, mouse,
   water, rocks

5. Desert ecosystem—sand, rattlesnake, cactus, river,
   jackrabbit

6. Tundra ecosystem—polar bear, seal, ice, pine tree,
   arctic fox

Mental Marvel Questions

Place the following parts of an ecosystem under the
correct heading of living or non-living.

	Living	Non-living
7. Pine tree		
8. Ocean sand		
9. Seagrass		
10. Pond water		

Think About It

What are three living and non-living things you might find in a rainforest ecosystem? If you need help, go to the websites listed below to find out more.

Be a Scientist

Have you ever played *I Spy* while driving in a car? This is a great game to play to look for ecosystems. Ask your mom or dad if you can take a walk around the neighborhood with them. Take along your journal, and use your *I Spy* skills to find ecosystems. Remember, they can be very small, so check everywhere. You might spot a corner that has a busy bunch of ants gathering food and building an anthill in the soft dirt. Or maybe you will be lucky enough to find some water that contains plants or animals. Draw every ecosystem you see, no matter how small it is. Be sure to label the living parts and the non-living parts of the ecosystem.

Food for Thought

Natural disasters such as fire and flooding are important for keeping ecosystems healthy. A forest fire can be helpful when it burns dead and sick trees making the entire forest healthier.

Coral is very tiny, and it takes many years to grow, but together, tiny coral can make reefs so big, they can be seen from space.

What's Online

Try these websites to find out more about ecosystems.

http://www.mbgnet.net/index.html

http://www.windows.ucar.edu/tour/link=/earth/rainforest. html&edu=elem

PLANTS AND ANIMALS HELP EACH OTHER

What You'll Find

TERMS AND DEFINITIONS

Pollination

Definition: Pollination helps a flower make seeds by moving pollen from one area of the flower to another.

In Context: Bees traveling from flower to flower help with pollination.

Dispersal

Definition: Dispersal is the act of spreading seeds in different directions.

In Context: Birds help plants with seed dispersal by eating fruit and spitting out the seeds.

Reproduce

Definition: Reproduce means to make new living things of the same kind.

In Context: Bees help flowers reproduce and make more flowers.

Your World and Science

Imagine you're lost and alone on a deserted island. What will you eat? Where will you live? Look around. What do you see? You see lots of plants, like tall, coconut-filled palm trees, leafy bushes bursting with berries, and thick, green grass. The plants will give you everything you need to live. The coconuts will give you sweet water to drink. The berries from the bushes will keep you from getting hungry. The green grass will give you the material you need to build a shelter to protect yourself from the sun. It will also give you a soft bed to sleep on at night.

Imagine you're back at home. Do you see plants around you that help you live every day? You stay healthy by eating fruits and vegetables. You stay cool and comfortable by wearing clothes made from plants like cotton. Without plants, surviving would be impossible, even if you weren't on a deserted island.

What You Need To Know

ANIMALS NEED PLANTS

Humans are not the only animals that need plants to live. All animals need plants for food and shelter. Remember the food chain from Chapter 2? We learned that all food chains start with plants. Herbivores would starve and die without plants. Without herbivores, carnivores would die too.

Animals need plants for shelter. Plants help animals hide and stay safe from predators. We know birds live in trees but so do bats, orangutans, and koalas. Bushes make wonderful homes for roadrunners, bobcats, and many types of lizards, while other animals, like the elf owl, Gila woodpecker, and cactus mouse enjoy the comforts of living in and around cacti. Without plants, many animals would be homeless, and they would be an easier dinner for predators.

Animals need plants for food, shelter, and even the air we breathe. But did you know that plants need animals too?

PLANTS NEED ANIMALS

Plants do not need animals for food or shelter, but they do need animals to help them reproduce. They need animals to help them make more plants. Plants need animals for pollination and seed dispersal.

POLLINATION

Plants make seeds through pollination. Some plants can make seeds by themselves, but most need the help of animals. Flowers use bright colors, strong smells, and sweet nectar to get the interest of bees, flies, and hummingbirds. The animals will feed on a flower's pollen or nectar. While it is feeding, it will get pollen stuck to its body or legs. Some of the pollen will rub off on the next flower it lands on. This is how pollination happens with the help of animals. Without pollination, plants could not make seeds. Without seeds, plants could not reproduce.

SEED DISPERSAL

Once the seeds are made, the plants need to get them to different places so they can grow. Without animals, plants would drop their seeds on the ground around them. The problem is there may not be enough nutrients or water for all the plants to grow. To give all the plants the best chance to live, plants need to spread to new places. Animals can help in many ways. They can carry the seeds on their fur. A fox walking through the forest might brush up against a plant that has sticky seeds. The seeds will stick to the fox's fur until it rubs up against something, and the seeds drop off far from where they started. An animal might eat a plant's fruit and leave the seeds on the ground in its poop. Many birds eat fruit and spit out the seeds. In this way, animals help plants with seed dispersal by spreading plants out in their habitats.

Let's Try It!

(Answer key page 203)

Einstein Questions

Can you remember what you read? Fill in the blanks. Use your highlighter to find key words in the text above.

1. _____ is the act of spreading seeds in different directions.

2. All animals need plants for _____ and
   _____.

3. Plants need animals for _____
   and_____.

Super-Genius Questions

Decide if the statement is **true** or **false**

4. All plants need animals for pollination. True or False

5. Pollination is how plants make seeds. True or False

6. Without plants, carnivores could still eat. True or False

59

Mental Marvel Questions

Choose the correct answer.

7. Animals are attracted to flowers because of
   _____.
   A. color and smell
   B. movement and location
   C. sunlight and water
   D. color and location

8. How can animals help plants with seed dispersal?
   A. They eat fruit and spit out the seeds.
   B. Seeds might stick to their fur.
   C. They pass seeds in their poop.
   D. All of the above

Think About It

Name two ways seed dispersal happens other than by animals. If you need help go to the websites listed below to find out more.

Be a Scientist

Become a real animal scientist. Go out to your backyard or neighborhood park and observe some pollination activity. Sit down quietly in a place where you can see flowers. What animals do you see around them? Record or list the insects, birds, and other animals you see. Don't forget to make a drawing of what you observe. What are they doing? Describe the activity taking place. Are the bees landing on different flowers? Are there hummingbirds flying around? **Record** all the **data** or information you find.

Food for Thought

The world's largest flower is called the corpse flower. It gives off a smell like a dead animal to attract flies for pollination. This is why it is called a *corpse* flower.

Animals need to be careful about what plants they choose to eat. While many are helpful, some are harmful like oak, buttercup, and mistletoe. All of these are poisonous.

What's Online

Try these websites to find out more about how plants and animals help each other.

http://library.thinkquest.org/3715/pollin5.html

http://www.mbgnet.net/bioplants/seed.html

Chapter 4

Energy

THE BIG IDEA

Energy has many forms and can be changed from one form to another.

ENERGY SOURCES

What You'll Find

TERMS AND DEFINITIONS

Energy

Definition: **Energy** is the ability to do work.

In Context: Food gives Joseph the **energy** he needs to run a mile.

Fossil Fuels

Definition: **Fossil fuels** are made in the Earth from dead plant or animal material.

In Context: Coal, oil, and natural gas are **fossil fuels**.

Renewable Energy

Definition: **Renewable energy** can't be used up.

In Context: Solar energy is **renewable energy** because the Sun can't be used up.

Your World and Science

What would happen if the Sun never came out? What if the world was always cold and dark, with no Sun to give it light and heat? Plants could not grow, which means there would be no food or oxygen. It also means there wouldn't be any energy. Not just food energy, but energy to run your parents' car or light up your house. It also means (you might want to sit down for this) no television or video games.

What You Need to Know

Energy comes to the Earth from the Sun in the form of light. When most people think of Sun energy, they think solar energy. But solar energy is not the only form of energy that comes from the Sun. Wind is also a form of energy the Sun helps to make. Millions of years ago, energy from the Sun helped to make fossil fuels like coal, oil, and natural gas. Now these fossil fuels provide energy for running cars, heating homes, and powering computers.

SOLAR ENERGY

Picture this—You're out in the middle of a hot, dry desert. There is nothing around for miles. All you can see are sun-baked rocks and hot sand. The few cacti that are growing here hardly give off enough shade to cool a lizard. What you want is an air-conditioned house and a tall, cold glass of lemonade. And then you see a shining white house out in the distance. This could be the answer to your wishes. But as you drag yourself across the sand, you see no power lines delivering electricity to this lonely place. Without electricity to cool it or to run a refrigerator, this house will be of no use to you. As you open the door expecting a terrible heat to come from inside, you get a blast of cold air instead. But how is that possible? There is no way to get energy out here. Or is there?

Solar energy is perfect for this desert environment. It is a clean, renewable source of energy that works well in sunny places like the desert. Before you can use energy directly from the Sun, it needs to be collected and changed into electrical energy by solar cells. Most of Earth's energy comes from the Sun, but very little of it is used as solar energy because you can't always depend on it. Solar energy works great in places like the desert, where the Sun shines most of the time. It is a clean source of energy because it does not pollute the environment. But areas with a lot of cloudy days would not collect enough sunlight to keep things running.

Wind Energy

Another clean, renewable source of energy that comes from the Sun is wind power. People have used the wind as a source of energy for thousands of years. They used the wind to pump water from the ground and grind corn into meal. The Sun warming the Earth's atmosphere makes wind. As the Sun warms the air around the Earth, some patches get warmer than others. As warmer air rises, cooler air blows in to replace it. This process is what causes wind. People use wind energy by putting up windmills. The wind turns the blades of the windmill and makes electricity. Wind energy is great for places that get a lot of wind, like along the beach. But the wind doesn't blow all the time, so we can't always count on wind power.

Stored Energy

Energy that comes from the Sun can also be stored for later use. The two most common forms of stored energy in nature are food and fossil fuel.

Food

In Chapter 2, we learned that plants use energy from the Sun to make food. We also learned that plants do not use all the food they make. Plants use some of the food energy to live and grow. The rest is stored in their leaves. When people and animals eat the plants, they can use the plant's stored energy.

Fossil Fuels

Most of the energy the world uses comes in the form of fossil fuels. These fuels, like coal, oil, and natural gas all come from the Sun. The Sun's energy is stored in plants and animals that died over 300 million years ago. To understand how fossil fuels were made you have to step back in time.

Picture this—You step into your time machine and dial up the Carboniferous Period over 300 million years ago. As you step out of your time machine, you enter a world very different from the one you know. It is a wet, swampy place covered with giant trees and huge, leafy ferns. You can see flying insects that are the size of your hand. You catch a glimpse of a lizard-like tail disappearing into the thick jungle. You decide this is not a good place to be. But before you go, you notice that as the plants die they sink down into the wet soil. Like today's plants, these prehistoric plants have gotten their energy through photosynthesis. The energy they have not used is now stored in their leaves. As you move forward in time, you can see the decomposing plants being covered with sand and mud. As millions of years pass by your time machine window, you observe that the mud is forming into rock. As the rock builds, it creates pressure and heat that squeezes all the water out of the plant material. This is the beginning of fossil fuel. In a few million years, this material will become coal.

Dead animal material also creates fossil fuels in the same way. Many scientists believe that diatoms, a tiny ocean animal no bigger than a pinhead, were the main source of oil and natural gas.

Burning fossil fuels is the only way to release the energy stored in them. The problem with fossil fuels is that they pollute the environment, and they take very long to make. They are not renewable energy sources. You don't need a time machine to know that once all the fossil fuels have been used, they are gone forever.

Careful!

Many people believe that fossil fuels were created from dead dinosaurs. Dinosaurs lived in the Cretaceous Period, 65 million years ago. Most of the fossil fuel deposits are from the Carboniferous Period, 360 million to 286 million years ago.

BATTERIES

We've talked about energy sources from nature, but humans can also store energy in the form of batteries. We use batteries everyday in things like TV remotes, cell phones, and cameras. The stored energy in batteries comes from chemicals. Batteries use the chemicals to make electrical energy. We will talk more about electrical energy in Chapter 7.

Let's Try It!

(Answer key page 203)

Einstein Questions

Can you remember what you read? Fill in the blanks. Use your highlighter to find key words in the text above.

1. _____ is the ability to do work.

2. Most of Earth's energy comes from the _____.

3. The Sun's energy is stored in both_____ and_____.

Super-Genius Questions

Choose whether the statement is true or false.

4. Fossil fuel is a form of stored energy. True or False

5. Dinosaurs are the major source of fossil fuel. True or False

6. Renewable energy is energy that is without limit. True or False

Mental Marvel Questions

Choose the correct answer.

7. Fossil fuel is made of _____.
   A. plants and animals that have been dead millions of years
   B. crushed up rocks and minerals
   C. super glue
   D. mud and dirt mixed together

8. Wind happens because as warm air rises _____.
   A. it creates air bubbles
   B. cooler air blows in to replace it
   C. it sucks air up with it
   D. the temperature goes up

Think About It

Why are renewable energy sources better for the planet than nonrenewable energy sources? If you need help go to the websites listed below to find out more.

Be a Scientist

Now you will have a chance to use the **scientific process** to see solar energy at work. You've done a lot of **observation** in the first three chapters, and now it is time to learn more about the **scientific process**. Scientists ask many **questions** before they begin an experiment. The **question** must fit the topic you are studying. You are studying solar energy in this experiment. There are many possible questions to ask about solar energy. Let's explore this **question**:

Question: How does the Sun's energy affect the movement of air?

The next step in the scientific process is writing a **hypothesis**. A **hypothesis** is your guess or a prediction about what will happen in the experiment. Read through the experiment and then come back to your science journal and write down a **hypothesis** of what you think will happen. The following questions might help you with your **hypothesis**. What do you think will happen to the balloons? Will they fill with air? Will only one fill with air? Which balloon will grow faster? Here is a good example of a **hypothesis** for this experiment:

Hypothesis: *I think neither balloon will fill with air.*

Now you need to get your **materials** together. Don't forget to ask your parents for help.

Materials: You will need two plastic bottles of the same size (without tops), white paint, black paint, paintbrushes, two small balloons, and your journal.

Procedures are the directions that tell you how to do the experiment. They need to be followed carefully. It is important to follow the **procedures** so that you get good results.

Procedures: Paint one bottle white and the other bottle black. After they have dried, fit the end of a balloon over the open mouth of a bottle. Repeat with the other bottle. Make sure the balloons fit tightly over the bottles. Place both bottles in direct sunlight. You should only have to wait a few minutes to see the results.

Record: What has happened? Touch both bottles. How do they feel? Which bottle is warmer? Did either of the balloons fill with air? **Record** exactly what happened in your journal.

Conclusion: To finish the experiment you need to write a **conclusion.** In your **conclusion** you should answer the question: "How does the Sun's energy affect the movement of air?" You also need to explain why your **hypothesis** was right or wrong.

Food for Thought

One minute of sunshine supplies enough energy to meet the Earth's needs for a whole year.

Natural gas has no smell, so it is mixed with a chemical to make it smell like rotten eggs. If you smell this in your home, get out. It might mean there is a gas leak.

What's Online

Try these websites to find out more about energy sources.

http://home.clara.net/darvill/altenerg/index.htm

http://www.eia.doe.gov/kids/energyfacts/index.html

MOVEMENT OF ENERGY

What You'll Find

TERMS AND DEFINITIONS

Wave

Definition: A **wave** is a disturbance that moves energy from one point to another.

In Context: The earthquake's energy made a big **wave** in the ocean.

Transfer

Definition: **Transfer** is to pass from place to place.

In Context: The teacher asked Jim to **transfer** his homework from his desk to her desk.

Your World and Science

It's time to get ready for bed. You walk into the bathroom, turn on the light, and start filling the tub for your bubble bath. You look into the mirror, see your reflection, and brush your teeth. Before you hop in the tub, you toss in your rubber ducky, and he bobs up and down in the moving bath water. As you sit down in the bubbles, you begin to sing a song. Your beautiful voice echoes loudly off the tile wall of the bathroom. There is a lot of energy in motion in this room.

What You Need to Know

MOVING OBJECTS

How does energy travel? One way energy travels is through moving objects. When you throw a ball to your friend, energy travels from you to the ball to your friend. If you bounce a ball, your energy is traveling through the ball. In this way, energy is moving from one place to another.

WAVES

Another way energy travels is through waves. Waves are energy in motion. Some examples of energy traveling in waves are water waves, sound waves, and light waves.

WATER WAVES

The waves you are probably most familiar with are water waves. Waves in the water move in an up and down motion. Waves in water can be caused by wind. The ocean is the best place to see waves caused by wind. Waves can be small or big depending on how much wind there is. The bigger the wind, the more energy there is to transfer to the water. Really big waves have enough energy in them to tip over a boat. The energy of an undersea earthquake can cause a huge wave called a tsunami or tidal wave. Sometimes humans can make waves when they are driving a boat.

LIGHT WAVES

Light is another form of energy that moves in waves. Like water waves, light waves move up and down. Light can travel through space, which is why we can see the light of the stars. Everything you see either gives off light or reflects light. The Sun gives off light energy in waves, but light bounces off the moon. We can see the moon because the Sun's light bounces off the moon's surface. Light waves cannot travel through solid objects. For example, when you go outside in sunlight, your shadow appears on the ground next to you. This is because light waves cannot travel through your body. As we learned in the last section, the energy from the Sun's light can be used for wind, food, and fossil fuels. It can also be used to produce electricity.

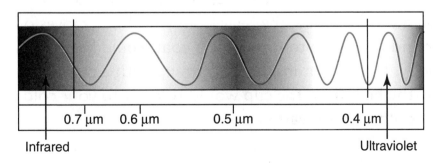

SOUND WAVES

Sound is energy that moves in waves. Sound waves vibrate and move back and forth like a coiled spring. When a sound is made, air is pushed, and it continues to push out in a vibrating wave until the energy is gone. When you speak, your vocal cords vibrate, and you hear your voice. The sound waves travel out in the direction you are facing, which is why your teacher will ask you to face her when you are talking to her. Unlike light waves, sound waves can travel through objects. This is why you can hear your father snoring in the next room. During a thunderstorm, a bolt of lightning heats the air quickly. The air expands making the sound of thunder. You see the lightning first because light waves travel much faster than sound waves.

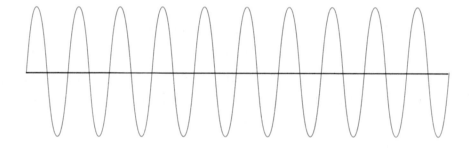

THERMAL ENERGY

Another way energy moves is through heat. Energy moves from hot to cold. If you hold an ice cube in your hand, the ice melts. The heat from your hand is moving into the cold ice, and it begins to melt. If you walk outside on a very cold day, the heat from your body moves out to the colder air and you begin to feel cold. A coat would keep the heat closer to your body. If you went inside to warm up with a cup of hot chocolate, the heat from the cup would transfer to your cold hands, and they would start to warm up.

ELECTRICAL CURRENT

Electricity is another way energy can move. Energy from the Sun can be used to make electricity, another type of energy in motion. Electricity is made by the free flow of electrons, which will be discussed in Chapter 7. Electricity is used to light your home, turn on your TV, and power your air conditioner on a hot day. People are very used to using electric power; life would be very hard without it.

Let's Try It!

(Answer key page 203)

Einstein Questions

Can you remember what you read? Fill in the blanks. Use your highlighter to find key words in the text above.

1. _____ are energy in motion.
2. Three examples of wave energy are_____,
   _____ , and_____.
3. _____ is to pass from place to place.

Super-Genius Questions

Decide if the sentence is **true** or **false**.

4. Light waves can travel through space. True or False
5. The cold from the snow moves into your hands when you make a snowball. True or False
6. Sound waves can move through a door. True or False

Mental Marvel Questions

Choose the correct answer.

7. Waves are _____.
   A. colorful planets
   B. energy in motion
   C. electrical current
   D. all of the above

8. A rainbow's colors are visible because of _____.
   A. sound waves
   B. water waves
   C. electrical current
   D. light waves

Think About It

Name four ways energy was in motion in the bathroom scene in "Your World and Science."

Be a Scientist

Take your science journal and a pencil and go outside to your backyard or a neighborhood park. Find a shady place to sit, close your eyes, and open your ears. What do you hear? Listen carefully, and try to **identify** the sounds around you. Do you hear kids talking, birds singing, cars driving by, or a dog barking? Open your eyes and write down what you have **observed**. Remember observation is not always with your eyes. Write down a lot of details.

Were the sounds loud or quiet? What directions were the sounds coming from? Were the sounds nearby or from far away? Were the sounds from nature, like a bird, or from a mechanical object, like a car? You can make categories or group the sounds you hear. For example, list all the soft sounds, and then list all the loud sounds.

Food for Thought

Light travels at the speed of 186,000 miles per second.

If you have ever seen a rainbow, you have seen light waves. Each color in the rainbow has a different wavelength. Red is the longest, and violet is the shortest. Together they make white light.

Sound waves cannot travel in space because there is no air in space. If a planet were to explode, you would not be able to hear it in space.

What's Online

Try these websites to find out more about the movement of energy.

http://www.miamisci.org/af/sln/

http://www.colorado.edu/physics/2000/index.pl?Type=TOC

SIMPLE MACHINES: CHANGING ENERGY TO MOTION

What You'll Find

TERMS AND DEFINITIONS

Motion

Definition: Motion is the act of changing place or position.

In Context: The motion of the roller coaster was making me dizzy.

Force

Definition: Force is a push or a pull that can change the movement of objects.

In Context: We had to force the heavy couch up the ramp to the house.

Friction

Definition: Friction happens when the surface of two objects rub against each other. It can help or hurt the movement of an object.

In Context: Good friction on the road kept our car tires from sliding out of control.

Gravity

Definition: Gravity is the force that pulls objects down to the ground.

In Context: If it weren't for gravity, we would all float around like astronauts on the moon.

Your World and Science

Imagine you are resting on the couch watching a scary vampire movie when your mom comes in and asks you to wash the windows. And from the look she gives you, she means now. Sadly, you turn off the TV and get off the couch. But before you go out in the hot sun, you decide to get a cold bottle of peach-flavored tea. You pull out a can opener and pop the top off. Next, you pull on the string that will open the blinds on the windows. With heavy feet you trudge down the back steps to the garage to grab a ladder. You put your bucket and cleaning supplies in a wagon and roll them to the back windows. With soapy sponge in hand, you climb the ladder and start washing the windows. Because you were so bummed about missing the final *bite* of the vampire movie, you did not notice that between the couch and the window, you used four simple machines. Simple machines are around us everyday. They make our lives much easier by changing energy to motion, but we hardly ever notice them.

What You Need to Know

Simple machines are all around us. When they are used to move things, they are showing their mechanical energy. Five of the most common simple machines are pulleys, levers, inclined planes, wheels and axles, and wedges. Get on your red tights and black mask. You are about to become a superhero. With the help of simple machines, you will be strong enough to move heavy objects.

PULLEY

You spend the morning watching your poor dad carry heavy cans of paint upstairs to paint the bedrooms. Suddenly, you have a superhero idea. You can use a pulley to make the job easier. A pulley has a wheel with a groove around the outside edge. A rope goes over the wheel and sits in the groove. The fixed pulley is attached to something overhead like a hook or a bolt in the ceiling. Pulleys make it much easier to lift a heavy object. You do not use as much energy because gravity makes it easier to pull down on the rope then to lift up on it. Now, instead of your dad lifting the heavy cans straight up the stairs, you attach the can of paint to one end of a rope that is thrown over a wheel and pull easily down on the other end. The paint can is lifted easily to the second floor. Your dad saves his energy, and you are a superhero.

LEVER

Feeling extra strong, you walk outside and find your little sister crying in the backyard. Instead of ignoring her like you usually do, you decide to find out what the problem is. It seems her scared kitten is stuck in a tree in the yard. The branch is too high for you to lift her up to, and you don't want to rip your tights by climbing the tree. You get another superhero idea. You drag your seesaw under the tree. The seesaw will act as a lever. When a straight bar or piece of wood is placed over a support point or fulcrum, it creates a lever. In this case, the fulcrum is the middle of your seesaw. A lever provides mechanical energy by allowing an object to be lifted using less force or effort. With a lever like a seesaw, you can use less energy by using less force to lift your sister up to the tree branch. It will take you less effort to lift your sister higher. Your superhero outfit will stay clean, the kitten will be safe, and your sister will stop her whining.

Inclined Plane

Now that your work is done in the backyard, you walk to the front of the house and spot your older brother trying to lift a heavy box filled with old junk into the back of his truck. You see that it is too heavy for him to lift, and you decide to come to his rescue with another superhero idea. An inclined plane is just what is needed. An inclined plane is a slanted flat surface. You grab your skateboard ramp and push it up to the back of his truck. An inclined plane can help you by making it easier to move objects up to a higher level. The ramp or inclined plane will make it easier to move the heavy box from the ground up to the truck. You will use less of your own energy. Of course you have to move the box for a longer distance, but it will take much less force to move the box.

Wheels and Axles

You have had a busy morning in your red tights, but your work is not done. Mom has pulled into the driveway with a van full of groceries. She looks tired. Superhero that you are, you know what she needs. The wheel and axle is a simple machine that has two parts, the wheel and the axle. The axle is a rod or small cylinder that goes through the middle of the wheel. It takes less force to turn the axle than the wheel. So when the axle is turned, the wheel begins to turn too. There is less friction created because only the outer surface of the wheel is touching the ground. You run and get your little sister's red wagon and pull it up to the van. Your mom is amazed as she

watches you put all the grocery bags into the wagon and roll them to the back door with little effort. It would be very difficult to pull a loaded wagon across your yard without the wheels and axles.

WEDGES

Your superhero day is not quite over. You are back upstairs to see if your dad needs help. He is trying to open the lid on the paint can. It is not an easy job. But never fear, you have another superhero idea. You hand your dad a flathead screwdriver and hammer. It's hard to believe that a screwdriver is a simple machine, but it is. The sharp edge on the flathead screwdriver is called a wedge. A wedge looks like two inclined planes stuck together in the shape of a triangle. A wedge helps you split things apart. The sharper the end of the wedge, the less force you need to split something apart. You place the wedge of the screwdriver between the lid and the can, and tell your dad to hit the end. The wedge pushes between the lid and the can, lifting the lid up. The can is open, and you have saved the day once again. Now it is time for a superhero nap.

Let's Try It!

(Answer key page 204)

Einstein Questions

Can you remember what you read? Fill in the blank. Use your highlighter to find key words in the text above.

1. A lever is a straight piece of wood placed over a support point or _____.

2. A flathead screwdriver is a good example of a _____.

3. A pulley has a wheel with a _____ on the outside edge.

Super-Genius Questions

Match the word with its definition.

4. Gravity   A. the force that pulls an object to the ground

5. Force   B. the force that happens when the surface of two objects rub together

6. Friction   C. the act of changing place or position

         D. a push or a pull that can move an

7. Motion   object

Mental Marvel Questions

Choose the correct answer.

8. Three examples of simple machines that lower the amount of force needed are _____.
   A. friction, gravity, inclined plane
   B. lever, inclined plane, wedge
   C. rope, flag, wedge
   D. knife, can opener, motion

9. When you are riding your bike uphill, you are using wheels and axles to ride up the _____.
   A. pulley
   B. friction
   C. lever
   D. inclined plane

Think About It

Go back to the "Your World and Science" section. Reread the paragraph, and find the four simple machines used on your way out to wash windows. If you need help go to the websites listed below to find out more.

Be a Scientist

Experiment with friction by building a simple ramp or inclined plane.

Question: *Does friction affect the movement of an object on an inclined plane?*

Hypothesis: Remember a **hypothesis** is your guess or prediction about what will happen in the experiment. Make sure to read through the experiment to understand what it is about before you write your **hypothesis** in your journal. The following questions might help you with your **hypothesis**. Will the sandpaper help or hurt the movement of the object on the inclined plane? Does friction make things move faster or slower? Here is a good example of a **hypothesis** for this experiment: *I think the object will move slower down the inclined plane with sandpaper.*

Materials: You will need ten paperback books, a smooth piece of cardboard, a sheet of sandpaper, a paper clip, and your journal.

Procedures: Make four stacks of book that line up, one stack behind the other. The first stack will have one book. The second stack will have two. The third stack will have three books, and the last stack will have four books. Next, lay the cardboard over the stacks so that the cardboard creates an inclined plane. Push the paperclip up and down the inclined plane. Now cover the ramp with a piece of sandpaper. How does the sandpaper affect the amount of force or energy you must use to push the paperclip up the ramp? Is it harder for the paperclip to slide down the ramp?

Record: What has happened? **Record** exactly what happened in your journal.

Conclusion: In your **conclusion**, you should answer the question: "Does friction affect the movement of an object on an inclined plane?" You also need to explain why your **hypothesis** was right or wrong.

Food for Thought

Some ancient cultures did not ever develop the wheel because they lived in rocky or steep land where a wheel would not be helpful.

Pyramid builders in ancient Egypt had to move heavy blocks of limestone up inclined planes to the tops of the pyramids.

What's Online

Try these websites to find out more about simple machines.

http://www.mikids.com/Smachines.htm

http://www.edheads.org/activities/simple-machines

Chapter 5

Matter

THE BIG IDEA

Matter has different forms and can be changed from one form to another.

SOLIDS, LIQUIDS, GASES

What You'll Find

TERMS AND DEFINITIONS

Matter

Definition: Matter is everything that takes up space.

In Context: The book that you are holding is made of matter.

Mass

Definition: Mass is the amount of matter in an object.

In Context: On the moon, you would weigh less, but your mass would be the same as on Earth.

Atom

Definition: An atom is the smallest piece of matter.

In Context: All matter is made up of atoms.

Molecules

Definition: Molecules are two or more atoms that are bound together.

In Context: One drop of water contains billions of water molecules.

Your World and Science

It's time for lunch. You're sitting at the kitchen table getting ready to eat the delicious lunch your mom has put in front of you. But before you eat, you take a minute to enjoy the warmth of the kitchen on a cold winter day. You notice that the table you are sitting at is hard and cool to the touch. The hot steam rising from the bowl of chicken noodle soup makes your face a little damp as you lean over it to get a good smell. Careful, you don't want to burn your mouth as you quickly spoon up some of the steaming soup. Everything in the kitchen is matter. Whether it is the solid table, the liquid soup, or the steam rising to dampen your face, matter is everything you see and feel. Even you and your mom are matter.

What You Need to Know

Matter is everything that takes up space and has mass. There are different states of matter. Most matter comes in the form of a solid, a liquid, or a gas. Matter is made of atoms. Atoms are very tiny particles or pieces. Atoms are always moving. They can move slowly or quickly. The higher the temperature the faster the atoms move.

Picture this—You are a molecule of water. Like all water molecules, you are made of two hydrogen atoms and one oxygen atom. You live in a bottle with your fellow water molecules. You do not live so close together that it feels crowded, but instead you move easily around, sliding past one another. Then one day your bottle is moved. It is much colder in this new place. In fact, it is freezing. It is no longer as easy to move around. Your fellow water molecules are all lined up now. They are hardly moving, but when they do move, it is all together. What is happening? You are still water, but you are no longer liquid. You are solid. You are ice, and you are not very happy about it.

Just as you start to get comfortable in your new state, something else happens. Your bottle is taken out of the cold. You slowly start to warm up. You and your friends begin to move around easily again. But before you can jump for joy, your bottle gets turned upside down. Out of the bottle you go, into a cold metal pan. You are stuck at the bottom of the pan unable to look out, but you can feel it getting hot. You and your friends are not staying together but moving farther and farther apart. You notice that your fellow water molecules are moving faster now. The hotter it gets, the faster they move. You're really flying past each other now. Then you realize you are no longer in the bottom of the pan. You are floating above it. You are still water, but you are not solid or liquid. You are a gas called steam. You are moving up and away from the other molecules. See you later, pals.

MATTER

Solid

Solid is the state of matter where molecules are packed closely together in neat rows. There is little space between the molecules, and the molecules cannot move past each other. Because the molecules are so closely packed, solids do not change shape. A desk is a solid. It does not change shape. You can't put your hand through it. Let's go back to the kitchen in "Your World and Science." What are some solids you see? The table and chair are solids. So are the bowl and spoon. Can you think of any more?

Liquid

Molecules in liquid are not as closely packed together as in solids. There is more space between each molecule making it easier for them to move. Because the molecules are able to move easily, liquid is able to change its shape. A liquid will take the shape of its container. If you were to fill a glass with water, the water would take the shape of the glass. Then if you were to take that glass of water and pour it into a fish bowl, the water would take the shape of the fish bowl. But you would still have the same amount of water as you had in the glass. So you might want to add more water if you are going to add a fish to the bowl. Let's go back to the kitchen. What are some of the liquids you see? The soup in your bowl is a liquid. Your mom might pour you a glass of cold milk with that soup, and milk is a liquid too.

Gas

Molecules in gas have a lot of space between them. Gas molecules fly quickly past each other. They are fast moving and do not stay close together. Molecules of gas will spread out to fill the container, no matter how big or small. If the container is open, gas will move out of the container. A good example of this is untying a balloon. Let's go back to the kitchen. What are some gases you see? The steam from your hot soup is water in the form of gas. There are also gases you can't see in the kitchen. The air you breathe is a gas too.

Let's Try It!

(Answer key page 204)

Einstein Questions

Can you remember what you read? Fill in the blanks. Use your highlighter to find key words in the text above.

1. _____ is everything that takes up space.

2. There is a lot of _____ between molecules in a gas.

3. _____ are two or more atoms bound together.

Super-Genius Questions

Match the state with its properties.

4. Gas   A. Molecules move past one another easily.

5. Liquid   B. Molecules are in neat rows.

6. Solids   C. Molecules move quickly and are far apart.

Mental Marvel Questions

Choose the correct answer.

7. Solids are _____.
   A. able to change their shape
   B. able to change the amount of space they fill
   C. hard to the touch and take up a certain amount of space
   D. all of the above

8. Three examples of liquids are _____.
   A. wood, shampoo, steam
   B. air, soup, milk
   C. rock, ocean, sand
   D. ketchup, soda, blood

Think About It

What are two other states of matter beside solids, liquids, and gases? If you need help, go to the websites listed below to find out more.

Be a Scientist

Now you will have a chance to identify different states of matter. For this activity, you will need your science journal and some old magazines. Ask your parents for magazines they no longer want because you will need to cut them up. Write the word "solid" on one page of your journal. On the next page write the word "liquid" and on a third page write the word "gas." Then look through the pages of the magazine to find pictures of matter in a solid state. Remember you are looking for things that have a definite shape like cars and buildings. Cut out the pictures you find and glue them on to the page marked *solid*. Find enough pictures to fill the page. Don't forget to label each picture. Repeat this activity for both liquid and gas. If you have trouble finding pictures, you can also draw some on the page but make sure to label what you draw.

Food for Thought

We think of metals as being hard and strong. We think of them as solids, but that's not always true. Metals like gallium or cesium have a melting point of 86 degrees Fahrenheit. They could melt on a really hot day. You certainly would not want to drive a car made out of these metals.

What's Online

Try these websites to find out more about solids, liquids, and gases.

http://www.chem4kids.com/files/matter_intro.html

http://www.chem.purdue.edu/gchelp/atoms/states.html

CHANGING MATTER

What You'll Find

TERMS AND DEFINITIONS

Evaporation

Definition: Evaporation happens when a substance changes into a gas.

In Context: Evaporation caused the rain puddle to disappear.

Condensation

Definition: Condensation causes a gas to turn into a liquid.

In Context: Condensation on the outside of the cold glass made it slippery.

Solution

Definition: Solutions are a mixture of two or more substances.

In Context: We mixed a solution of yellow and blue paint to make green.

Your World and Science

You are sitting in your grandmother's yard enjoying a frozen grape ice pop when your baby brother calls you over to give him a push on the swing. You lay the purple ice pop down on its wrapper and run over to give the little guy a push. He is having so much fun you decide to keep pushing a few more times. You run back to your grape ice pop only to find that it is now a purple puddle with a stick floating in the middle. You are sad until your grandma comes out and turns on the sprinklers. You run through the cool water forgetting all about your melted grape ice pop. A while later you come back to the place where you left the purple puddle of melted ice, and all that is left is a sticky layer of purple goo. Your grape ice pop went through two big changes while you were playing in the yard. As you learned in the section above, most matter is solid, liquid, or gas. But as you saw with the ice pop, matter can change from one form to another.

What You Need to Know

You have learned that matter can be solid, liquid, or gas. But matter can change forms with the help of heat and cold. A solid like the grape ice pop can melt into a liquid when placed in the hot sun. The hot sun can then change the water from the melted ice pop into a gas. If you placed the melted ice pop into a freezer, it would be a frozen ice pop again.

PHYSICAL CHANGES

Melting

If you have ever melted a piece of ice in your mouth, you have caused a solid to change to a liquid. The heat inside your mouth made the slow-moving molecules in the solid ice move faster. Every solid has a temperature where it will melt. This is called the melting point. The melting point is the temperature a solid must reach to become a liquid. Solids have different melting points. For example, it would take a much higher temperature to melt a piece of plastic or metal than an ice cube. While you could easily melt an ice cube in your mouth, you would not be able to melt a straw or a paper clip with the heat in your mouth.

Freezing

Just as solids have a melting point for turning into liquid, liquids have a freezing point for turning into a solid. When you fill an ice cube tray with water and place it into the freezer, the molecules in the water slow down in the cold air. When the temperature of the water reaches its freezing point, the molecules have slowed down enough to become a solid piece of ice. Not every liquid freezes at the same temperature. Grape juice takes longer to become a solid than water does.

Evaporation

If you have ever watched a pot of water boiling on the stove, you have watched evaporation in action. Evaporation happens when a liquid is heated and becomes a gas. The water in the pot on the stove has moving molecules. As the water heats up, the molecules move faster.

Soon some of the heated water molecules escape the pot and float into the air. These water molecules are now a gas called water vapor. Evaporation does not always need heat. Liquids all have some fast-moving molecules that can escape and become a gas without the help of heat. After a rainstorm, a puddle will slowly evaporate even if the temperature outside remains cold, and the sky is still cloudy.

Condensation

Condensation happens when a gas turns back into a liquid. After you have taken a long, hot bath or shower, you notice that the mirror over your sink is covered with a wet fog. The walls of the bathroom are also wet. While you were bathing, some of the molecules of hot water escaped the shower and floated up as the gas we call water vapor. When the gas molecules of the water landed on the cold surface of the mirror and walls, the molecules cooled off and turned back into liquid water. This is called condensation.

Mixtures

Some matter may be combined with other matter to form a mixture. Physical forces like heat, stirring, or shaking can hold together different substances. For example, if you put paper clips and marbles into a bowl and stirred them around, you would have a mixture. But you could still easily separate the marbles from the paper clips. You can separate the matter from the mixture, and the molecules have not changed. When you pour a big spoonful of chocolate powder into a tall glass of milk, you are making a mixture. You help mix the solid powder and liquid milk together by stirring it with your spoon. And if you don't stir it well enough, you find a big lump of wet chocolate powder on the bottom of the glass after the milk is gone. And, of course, this is the best part of a glass of chocolate milk. With a little effort and some scientific equipment, you could separate the chocolate powder from the liquid. But then you wouldn't have any chocolate milk to drink. The important thing to remember about mixtures is that the molecules of each substance do not change.

Solutions

When a mixture of two substances like sugar and water is heated, the sugar dissolves in the water. The sugar is then evenly dissolved throughout the glass of water. This makes the sugar water a solution. When the amount of sugar molecules is the same and there is not more sugar in one part of the water than another, this is called a solution. You can also separate the sugar from the water by allowing the water to evaporate. After the water evaporated, only the sugar would be left. The grape ice pop you read about in "Your World and Science" was an example of a solution. After the purple puddle sat in the sun for a long time, the water from the solution evaporated. All that was left behind was the grape goo used to make the ice pop.

Careful!

Solutions are also mixtures. But solutions have molecules that are evenly divided throughout the mixture. Do you remember the chocolate milk example? That was a mixture because you had more chocolate powder at the bottom of the milk than the top. But the grape ice pop was a solution because the sugar and flavor molecules were evenly mixed throughout the whole ice pop making sure that every lick was as sweet as the last.

Chemical Changes

When a molecule has changed and it is no longer the same molecule, then it has gone through a chemical change. When metal rusts, the molecules of metal have changed and turned to rust. When you burn a piece of wood in the fireplace, the log of wood disappears, and there is a pile of soft, gray ash left behind. The molecules in the wood have changed and become ash.

Let's Try It!

(Answer key page 204)

Einstein Questions

Can you remember what you read? Fill in the blanks. Use your highlighter to find key words in the text above.

1. Evaporation happens when a liquid becomes a
   _____.

2. Matter can change forms with the help of
   _____ and _____.

3. A liquid becomes a solid when it reaches its
   _____.

Super-Genius Questions

Match the change of matter to the physical change.

4. Boiling water to water vapor   A. evaporation

5. Orange juice to ice pop   B. melting

6. Snowball to puddle   C. freezing

Mental Marvel Questions

Choose the correct answer.

7. An example of molecules that have gone through a
   chemical change is _____.
   A. cereal mixed with milk
   B. water turning to ice
   C. burnt toast
   D. hot water turning to steam

8. Your bathroom walls are wet after a hot shower
   because of _____.
   A. condensation
   B. melting
   C. freezing
   D. chemical changes

Think About It

What temperature is the freezing point for ice? At what
temperature does water boil? If you need help go to the
websites listed below to find out more.

Be a Scientist

Making ice cream is the tastiest way to change a liquid to
a solid. You will need some help with this. In fact the
whole family might enjoy this activity. To make ice cream
you will need: ¼ cup of milk, ¼ cup of heavy whipping
cream, ⅛ cup of sugar, and ⅛ teaspoon of vanilla. You will
also need a large Ziploc® bag and a small Ziploc® bag,

ice, and ½ cup rock salt. Pour the milk, cream, sugar, and vanilla into the small Ziploc® bag and seal it tightly. Put ½ cup of rock salt and some ice cubes into the large Ziploc bag. Now place the mixture into the big bag and seal it tightly. Shake the bag until the mixture turns into a solid that looks like ice cream. And now for the best part, grab a spoon and eat it! It's important to remember that the rock salt was added to lower the freezing point of the ice to keep it frozen longer.

Food for Thought

Some matter can go straight from being a solid into a gas form in a process called sublimation. Dry ice is frozen or solid carbon dioxide. When it warms up, instead of melting, dry ice turns into a gas. That is the foggy white smoke you see at Halloween parties.

Planet Earth is covered with solutions we call oceans.

What's Online

Try these websites to find out more about changing matter.

http://www.bgfl.org/bgfl/custom/resources_ftp/client_ftp/ks3/science/changing_matter/index.htm

http://www.bbc.co.uk/schools/scienceclips/ages/9_10/science_9_10.shtml

Chapter 6

Light

THE BIG IDEA

Light travels and is reflected and absorbed by objects.

REFLECTION AND ABSORPTION

What You'll Find

TERMS AND DEFINITIONS

Reflection

Definition: **Reflection** happens when light waves bounce off a surface.

In Context: My dog was barking at his own **reflection** in the mirror.

Absorption

Definition: **Absorption** is the process of taking in something.

In Context: The **absorption** of heat into the frying pan handle made me burn my hand.

Transparent

Definition: If an object is **transparent,** you can see through it.

In Context: I looked through the **transparent** glass in the window.

Opaque

Definition: An **opaque** object does not allow light to travel through it.

In Context: You cannot see through the wall in your bedroom because it is **opaque.**

Your World and Science

You're sleeping in your cozy bed in your dark room and thunder wakes you from your dreams. You stumble out of bed and smack your toe on a chair. Ouch! You reach for the light switch and gasp when you find that the power is off. The storm must have knocked out the electricity. Then a white flash of light blinds you for a second, and the room goes dark again. You take another step and trip over the pair of shoes you left lying in the middle of your room. Now you've learned two things. You've learned that you need to clean your room, and you've learned that you cannot see the mess in your room without light. The objects are there even in the dark. But without light, they are hard to see. This chapter will explain how light travels and how it lets you see objects and colors.

What You Need to Know

SHADOWS

Whether light comes from the Sun, the lamp in your room, or the lightning from a storm, it travels in a straight line. If you turn on a flashlight and then place your hand in front of the light, a shadow of your hand will appear on the wall. The light traveling from the flashlight cannot travel through your hand. While light still appears on the wall, the area where your hand blocks the light has made a shadow. When you move your hand or fingers, the shadow moves to show where the light is being blocked. When you are walking on a sidewalk in the sunlight, you can see your shadow on the white sidewalk. You are blocking the sunlight with your body.

REFLECTION

As light hits an object, some of the light is reflected or bounced off the object. The light travels back to your eye and lets you to see the object. You can also see yourself in a mirror because light bouncing off the mirror is reflected back into your eyes.

Have you every stood on the edge of a lake and looked at your own reflection in the water? When light bounces off a smooth surface like a lake or a mirror, the reflection is clear. You can see exactly what is reflected. But if a strong wind blew across the smooth lake, it would be hard to see your reflection clearly in the wavy water. The light would be bouncing off in all directions, and your reflection would be blurry.

Some objects are transparent. These objects do not reflect much light. Instead, the light passes through the object. Clear glass, air, and water are mostly transparent. Light passes through them.

But many objects are not transparent, and they block light. When an object does not allow light to travel through it, like a wall or your hand, it is called opaque.

ABSORPTION

Most objects reflect light. Light that bounces off the object helps us to see them. These objects also absorb some of the light that hits them. This means that some of the light energy is taken into the object. While your hand would stop the light from a flashlight, if you left the light shining on it long enough, your hand would begin to feel warm. It would be absorbing some of the light energy from the flashlight.

Have you ever wondered why a pumpkin looks orange or a banana looks yellow? The light we see around us is called white light because it is made up of all the colors. When you look at a tomato, you see the color red. The tomato has absorbed all the light except the red, which is reflected back to your eye.

Let's Try It!

(Answer key page 205)

Einstein Questions

Can you remember what you read? Fill in the blanks. Use your highlighter to find key words in the text above.

1. Light travels in a _____ line.

2. Light passes through _____ objects.

3. A _____ may appear when light is being blocked by an opaque object.

Super-Genius Questions

Decide if the statement is **true** or **false**.

4. Air is opaque because light travels through it. True or False

5. You can see your reflection clearly in the ocean waves. True or False

6. White light is made up of all the colors. True or False

Mental Marvel Questions

Choose the correct answer.

7. You see your reflection in the mirror because

   _____.

   A. you always look great
   B. light bounces off the mirror and back to your eyes
   C. white light is produced inside the mirror glass
   D. the mirror is transparent

8. A tomato looks red to you because _____.
   A. the tomato absorbs all the colors of light except the red
   B. the tomato has chlorophyll inside it
   C. the seeds make the tomato look red
   D. the sun has burned the skin on the tomato

Think About It

How is a rainbow formed? If you need help go to the websites listed below to find out more.

Be a Scientist

You can use light and reflection to write some secret messages. In your journal, write a secret message to a friend, but write it backwards. It might be easier to write your message on a separate piece of paper and then copy it backward in your journal. This is a tough one because even the letters need to be backward. Give your message to a friend and have him (or her) hold it up in front of a mirror. How does the message look in the mirror? Is it easy to read? How does the mirror help your friend read the message?

Food for Thought

You can never reach the end of a rainbow because the raindrops reflecting the light in the rainbow are always in a different part of the atmosphere.

What's Online?

Try these websites to find out more about absorption and reflection.

http://acept.la.asu.edu/PiN/mod/light/reflection/ pattLight1.html

http://www.deltatech.com/rv/rainbows.html

REFRACTION

What You'll Find

TERMS AND DEFINITIONS

Refraction

Definition: Refraction is the change of direction of a beam of light.

In Context: The ball sitting at the bottom of the pool looked much closer to the surface because of light refraction.

Your World and Science

Your little brother is bothering you again. You are trying to do your homework, but he is playing with his toy soldiers all around you. After he makes his soldiers run over your math paper for the third time, you've had enough. You grab up a handful of soldiers and throw them across the room. One of the soldiers lands in a glass of water. Good shot! Suddenly your brother starts screaming that you broke one of his soldiers. From where you're standing, it does look like the soldier's head was cut off. Your mom has heard all the noise your brother is making. She wants to know what is going on, and your brother tells her that you broke his toy for no reason. He wasn't doing anything. Your mom is already mad that you threw toys in the house, but before you admit to breaking anything, you take a closer look at the toy in the glass. As you get closer, you realize you are saved. You lift the toy out of the glass, and, like magic, the toy is in one piece. You are so glad you listened in science class because you have just been saved by light refraction.

What You Need to Know

As you learned in the last section, light travels in a straight line. An object can reflect or absorb light. You also learned that light can pass through a transparent object. Refraction happens when light goes through a transparent object. Light goes through the glass of a window, water in a lake, and the air. As light travels through these things, it travels at different speeds. Light travels faster through air than through glass or water. Light will move slower through objects with more closely packed molecules. In Chapter 5, we learned that the molecules in solids were packed closer together. So light does not travel as quickly through a solid as it does a liquid or gas. So when light travels from air to water, it changes speed and causes the light to slow down. But if the light goes from air to water at an angle, it will cause the light to refract or bend. Light bends when it changes speed and direction. If you put a pencil in a glass of water at an angle, it will look as if the pencil is broken like the toy soldier in "Your World and Science."

Picture this—You and your sister are at the grocery store with your mom. You both want to push the shopping cart. Your mom says you can both push the cart. You start out just fine, both of you pushing at the same speed, but then you see the last box of your favorite cereal on the shelf. You start pushing the cart faster than your sister. The cart turns, and you run into the grocery shelf. You changed speed and your sister didn't, causing the shopping cart to turn or bend. The same thing happens to light. Refraction is the bending of light that happens when light changes speed.

Let's Try It!

(Answer key page 205)

Einstein Questions

Can you remember what you read? Fill in the blanks. Use your highlighter to find key words in the text above.

1. Light can be either _____ or absorbed.

2. _____ is the bending of light.

3. Light changes _____ when it travels from air to water.

Super-Genius Questions

Decide if the statement is true or false

4. Light only bends if it changes speed at an angle. True or False

5. Light moves faster when it moves from air to water. True or False

6. The more closely the molecules are packed in an object, the slower the speed of light. True or False

Mental Marvel Questions

Choose the correct answer.

7. Refraction is the _____.
   A. absorption of light
   B. distance light travels
   C. bending of light
   D. all of the above

8. Three examples of transparent objects are _____.
   A. window, pool, air
   B. air, wood, milk
   C. rock, ocean, wall
   D. water, rubber, cotton

Think About It

What is the difference between reflection and refraction? If you need help, go to the websites listed below to find out more.

Be a Scientist

Now it's your turn to experiment with bending light.

Question: *Does light bend or refract when moving from air to water?*

Hypothesis: Remember a hypothesis is your guess or prediction about what will happen in the experiment. Make sure to read through the experiment to understand it before you write the hypothesis in your journal.

Materials: You will need a clear drinking glass, water, a pencil, and your journal.

Procedures: Fill the glass about half full of water. Then put the pencil into the glass. Lean the pencil against the side of the glass. Describe how the pencil looks. Fill the glass with more water. Describe how the pencil looks.

Record: What has happened? Record exactly what happened in your journal.

Conclusion: In your conclusion you should answer the question: "Does light bend or refract when moving from air to water?" You also need to explain why your hypothesis was right or wrong.

Food for Thought

The reason that we don't notice refraction when looking through a window is because light goes through both the inside and outside surface of a window. Light is bent one way when going through the inside surface of a window and then the other way when going through the outside surface of a window.

What's Online

Try these websites to find out more about refraction.

http://www.creativekidsathome.com/kids_science.html

http://www.opticalres.com/optics_for_kids/kidoptx_p1.html

Chapter 7

Electricity and Magnetism

THE BIG IDEA

Electricity and magnetism are closely related and are used in everyday life.

ELECTROMAGNETS

What You'll Find

TERMS AND DEFINITIONS

Attract

Definition: **Attract** is to pull with physical force.

In Context: A magnet will **attract** a metal paper clip.

Repel

Definition: **Repel** is to push away with force.

In Context: Magnets will **repel** or push away if the same poles are facing each other.

Electron

Definition: An **electron** is a negatively charged particle that floats around the nucleus of an atom.

In Context: Hydrogen has only one **electron**.

Current

Definition: **Current** is the movement of electric charge.

In Context: When the electric **current** was turned off, the light went out.

Your World and Science

The school bell rings at the end of the school day. You walk out of class looking for your ride home. Your mom pulls up and you jump in the car, happy for your afternoon of freedom. On the way home, you think about the shows you'll watch on the television and the music you'll listen to on the radio. But first you're going to get the blender out to make that chocolate shake you've been dreaming about all day. As soon as the car stops at your house, you reach for the door ready to put your plan into action. That is when reality hits. Your mom says to grab some fruit for a snack before starting your homework. And remember no television or radio until homework is done. Oh well, the television and radio can wait, but it sure would have been nice to have that chocolate shake. An apple just isn't the same. In this section, you will learn how this whole scene would not be possible without the help of electromagnets.

What You Need to Know

MAGNETS

Picture this—You and your family are on your way to a camping trip. Your mom bought you a board with magnetic letters to keep you from fighting with your brother. The letters are boring, but who doesn't like playing with magnets? Two ends stick together, flip one end, and they spring apart. Of course you know this is because magnets have two poles, a north pole and a south pole. If you put a north pole of one magnet next to the south pole of another magnet, they will be attracted to each other because opposites attract. But if you put a north pole of one magnet next to the north pole of another magnet, they will be repelled by each other. Although you are having a great time watching your magnets attract and repel each other, it would be more fun to use your magnets to drag the chain of paper clips out of your brother's pocket without his knowing it. Magnets attract certain metals like steel, iron, and nickel. So the paper clips should not be a problem.

MAGNETIC FIELDS

In order for the paper clip mission to be a success, you have to get close enough for the paper clips to be in your magnet's magnetic field. This is the area that surrounds a magnet at all times. A magnet can only repel or attract metal objects in its magnetic field. The field is strongest at the magnet's poles. The field lines are circular and travel from the north pole to the south pole and back again in a closed loop. A magnetic field is caused by the motion of negatively charged spinning electrons in the magnet. Silently, you push the magnet toward your brother's pocket. He doesn't have a clue about your plans. Just a little closer, and the paper clips will be yours. Too late! He's discovered your evil plot. You'll have to find another way to pass the time.

COMPASS

You're walking through the forest with your brother, and a smelly skunk scares you off the path. Your brother is worried that you are lost, but you know exactly where you are thanks to your Scout compass. A compass is a simple device used to find magnetic fields. A compass is made with a small, light magnet placed or balanced on a point like the spinner of a board game. The magnet is usually called a needle. One end of the needle is marked or painted to show that it points north.

Your brother is not impressed with your compass. He thinks if it only points north, it must be broken. You try to explain the Earth is surrounded by a magnetic field. The field is very weak because it comes from the center or the core of the planet. Earth is one big magnet with two poles just like any other magnet. The south pole of the Earth's magnet is located near the place we call the North Pole. The north pole of the magnet is located near the place we call the South Pole. Since opposites attract, the needle of the compass will always point north. Of course, your brother stopped listening after you mentioned the North Pole because he was too busy dreaming about flying reindeer.

ELECTROMAGNETS

Returning to camp, you see your dad searching in the grass by the lake. He asks you to help him pick up all the fishhooks he's dropped in the tall grass. You are not excited about spending your time picking up fishhooks. Then you get an idea. Fishhooks are made of metal and should be easy to pick up with a magnet. But the little magnets in your letters are not big enough to do the job. Another great idea pops into your head. You could make a magnet strong enough for the job. All you need is an iron nail from your dad's tool kit, wire long enough to wrap around that nail, and a battery from the flashlight. This type of magnet created with electrical current is called an electromagnet. Unlike a regular magnet whose magnetic field always surrounds it, an electromagnet's field only works when it is connected to a battery.

Electromagnets work because of the small magnetic field created by the electrons running through the wire. If you look closely at a D battery, you will see that there is a positive end with a + symbol and a negative end with a − symbol on it. Electrons flow from the negative end of the battery to the positive end. As the electrons flow in the wire, a magnetic field is created. If you wrap wire around a nail, a magnetic field will surround the nail. The nail will act like a magnet with a north pole and a south pole when attached to a battery. This electromagnet will pick up the fishhooks and impress your father at the same time.

The electromagnet does more than pick up fishhooks. Without it, you would not have the telephone to call your friends or a computer to surf the net. In fact, without the electromagnet, there would not be any motors—not small ones that run your blender or big ones that move your parent's car. An electric motor is basically a rotating electromagnet built within a non-moving magnet to make current. This type of motor is used in all the electrical appliances in your home as well as the generator you and your family use on a camping trip.

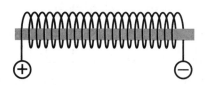

Let's Try It!

(Answer key page 205)

Einstein Questions

Can you remember what you read? Fill in the blanks. Use your highlighter to find key words in the text above.

1. The opposite ends of two magnets _____ each other.

2. Magnets only repel and attract objects in their _____.

3. An electromagnet needs a _____ to work.

Super-Genius Questions

Choose whether the statement is **true** or **false**.

4. The Earth has a very strong magnetic field. True or False

5. The Earth's magnetic south pole is near the place we call the North Pole. True or False

6. Your parents' car would not move without an electromagnet. True or False

Mental Marvel Questions

Choose the correct answer.

7. A magnet will attract _____.
   A. steel, iron, and nickel
   B. iron, plastic, and aluminum
   C. nickel, copper, and salt
   D. rubber, Styrofoam, and water

8. A magnetic field is created by _____.
   A. the magnet's motion
   B. flying reindeer
   C. the needle of a compass
   D. the movement of electrons

Think About It

Look back at "Your World And Science." Find the five things that work with the use of an electromagnet.

Be a Scientist

Let's explore this **question**:

Question: *What objects can be used to make the core of an electromagnet?*

Next in the scientific process is writing a **hypothesis**. A **hypothesis** is your guess or a prediction about what will happen in the experiment. Here is a good example of a **hypothesis** for this experiment:

Hypothesis: *I think only iron will work as a core to make an electromagnet.*

Now you need to get your materials together. Don't forget to ask your parents for help.

Materials: You will need an iron nail, a plastic straw, some wire (thin copper wire if possible), a D battery, paper clips, and your journal.

Procedures are the directions that tell you how to do the experiment. They need to be followed carefully. It is important to follow the **procedures** so that you get good results.

Procedures: Neatly wrap the wire around the iron nail being sure to leave enough wire on either side of the nail to attach to the battery. The more wire you wrap around the nail, the stronger your magnet. Once you are finished wrapping the nail, attach the wire to the battery. One end of the wire should be attached to the positive terminal of the D battery, and the other end of the wire should be attached to the negative end of the D battery. Now test your electromagnet by trying to pick up some paper clips. Repeat the above procedure with the plastic straw as a core.

Record: What has happened? How did each of the electromagnets work? Which one was strongest? Did they both work? **Record** exactly what happened in your journal.

Conclusion: To finish the experiment, you need to write a **conclusion**. In your **conclusion** you should answer the question: "What objects can be used to make an electromagnet?" You also need to explain why your **hypothesis** was right or wrong.

Food for Thought

A compass is not the only thing that can detect a magnetic field. Many animals can too. Pigeons, honeybees, and dolphins are some animals that can detect the Earth's magnetic field. They can use it to find their way.

Trains in Japan are made to float above the tracks with the help of magnets. Maglev trains use powerful magnets on the tracks and the underside of the train. The magnets repel each other, allowing the trains to travel up to 300 miles per hour above the track.

What's Online

Try these websites to find out more about electromagnets.

http://science.howstuffworks.com/electromagnet2.htm

http://www.exploratorium.edu/snacks/iconmagnetism.html

ELECTRICAL CHARGES AND ENERGY

What You'll Find

TERMS AND DEFINITIONS

Nucleus

Definition: The nucleus is the center part of something.

In Context: The nucleus is in the middle of an atom.

Protons

Definition : Protons are positively charged particles inside the nucleus of every atom.

In Context: The nucleus in an atom of carbon contains six positively charged protons.

> ### Neutrons
>
> *Definition* : The nucleus of an atom has particles with no electrical charge called **neutrons**.
>
> *In Context*: The **neutrons** found inside the nucleus of an atom have no charge.

Your World and Science

You wake up to a beautiful fall morning. The air outside is a bit chilly, and a very dry breeze is blowing through the colorful leaves on the trees. You smile because you remember it is picture day at school, and you have a cool, blue ski sweater to wear for the class photo.

You happily head to the kitchen for breakfast not aware that trouble is waiting for you in the dry morning air. You scoot quickly across the hall carpet in your warm wool socks and reach down to pat your baby brother on the head as you enter the kitchen and zap! You shocked your brother on his bald head with a small electric shock. He starts to cry, and your mom yells at you for upsetting your brother. You say, "sorry!" and sit down to your bowl of rainbow-colored cereal.

Once you gobble up your breakfast, you head upstairs to pull your sweater out of the laundry basket and comb your hair. As your head pops through the neck hole of your sweater, you look in the mirror and see that you hair is now stuck to your head and face. You grab a plastic comb and run it through your hair. Your hair sticks up in the air. You look terrible, and it's picture day! With no time to waste, you grab your backpack and head to school. But once you are in class and you've hung up your backpack, you notice that everyone is pointing at you and giggling. Your best friend whispers in your ear that you have something stuck to the back of your sweater. You reach back and pull a pair of your brother's Superman underwear off the back of your sweater. Your perfectly nice day has been ruined by static electricity.

What You Need to Know

STATIC ELECTRICITY

Everyone has felt static electricity. It might happen when you touch a doorknob, your dog's head, or your friend's arm. But how and why does it happen? As you learned in Chapter 5, all matter is made up of tiny parts called atoms. Each atom has a center or nucleus. The nucleus is made up of positively charged particles called protons. The nucleus also contains particles called neutrons that have no charge. Circling or orbiting around the nucleus of the atom are particles with a negative charge called electrons. Confused? Maybe it would be easier to understand if you were the size of an atom.

Let's take a ride on an electron. This particular electron is orbiting around the nucleus of a carbon atom like the Earth orbits around the Sun. You notice as you are spinning around on your electron that there are five other electrons whizzing around with you. Thanks to your first grade math teacher, you quickly add the numbers and find that there are six electrons in a carbon atom. You are a little dizzy, but you notice that there are 12 particles tightly bunched together in the center of the moving electrons. Thanks to your science teacher, you know that six of the

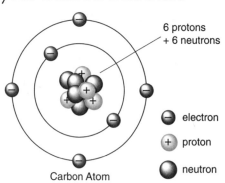

6 protons
+ 6 neutrons

electron
proton
neutron

Carbon Atom

particles are protons with a positive charge, and six are neutrons with no charge. Like the carbon atom, most types of atoms have the same number of positively charged protons and negatively charged electrons. So there is a neutral charge on the atom, which means no electrical charge is produced.

Suddenly, you are pushed around, and the electron you are riding on jumps to another atom. Now this other atom has more electrons than protons. Because it now has more electrons than protons, it has become negatively charged. Now you are floating in negatively charged particles. This is static electricity.

Let's bring you back to normal size and take you to the fair. You spot a very cool giant balloon and decide to buy it. When you hold the balloon in your hand, you do not feel any static electricity. But when you rub the balloon over your sister's hair and pull the balloon away, her hair sticks up in the air. If you listen closely, you can even hear the crackle of electricity between the balloon and her hair. She wonders why everyone is staring at her, but you think it best not to mention the strange hairdo.

Why is her hair standing up? As you rubbed the balloon on her hair, negatively charged electrons moved from the atoms in her hair to the balloon. Only the electrons moved, leaving her hair positively charged because there were now more protons than electrons in her hair. But that still doesn't explain why. Remember from the last section that the same poles of a magnet repel each other? The same electrical charges also repel each other. The strands of your sister's hair have the same electrical charge, and they are trying to stay far away from each other. The charge built up in your sister's hair is static electricity because it is not moving.

Electric Current

Static electricity is electricity that has not moved. The extra electrons in a negatively charged object are just waiting to move to a positively or neutrally charged object like a doorknob. When they move, an electric charge is released. The static electricity becomes electric current.

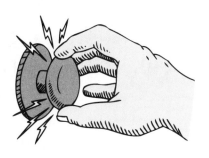

The crackle or snap you hear when you touch a doorknob and the small shock you feel in your hand come from the movement of electric charges. Rubbing your feet on a carpet causes negatively charged electrons to move from the carpet to your body. When you touch the doorknob, the electric charge moves to the doorknob. When electric charges flow from one object to another it is called an electric current. When electric current has a path to follow, it can help us in everyday life by giving us heat, light, and even sound.

Let's Try It!

(Answer key page 206)

Einstein Questions

Can you remember what you read? Fill in the blanks. Use your highlighter to find key words in the text above.

1. The nucleus of an atom is made up of _____ and _____.

2. An electron has a _____ charge.

3. Most atoms have _____ numbers of electrons and protons.

Super-Genius Questions

Match the atomic particle to its charge.

4. Neutron   A. positive charge

5. Proton   B. no charge

6. Electron   C. negative charge

Mental Marvel Questions

Choose the correct answer.

7. Electrical charge is caused by moving _____.
   A. protons
   B. atoms
   C. electrons
   D. neutrons

8. Static electricity is _____.
   A. a charge that has not moved
   B. charges that move between objects
   C. balloons at a fair
   D. moving electric current

Think About It

Look back at "Your World and Science." What three events were examples of static electricity?

Be a Scientist

Use static electricity to decorate yourself with balloons. Blow up 10 to 20 small balloons and tie off the ends. Rub the balloons, one at a time, on a piece of carpet or a wool sweater, and then stick the balloons on your clothing. How many balloons can you get to stick to your clothing at once? Try different types of clothing. Does a cotton t-shirt work as well as a wool sweater for holding on to the balloon? Try a nylon sports jacket or a piece of clothing that contains polyester. Which clothing is the best for holding onto the balloons? Would the balloons stick better if you rubbed them on the carpet for a longer time? How long did the balloons stick? What do you think makes the balloons fall off eventually? Describe the results in your science journal.

Food for Thought

Static electricity can be used to move a stream of water. If you hold a negatively charged piece of plastic near a flowing stream of neutral water, the water will be attracted to the plastic and bend toward it.

Electricity can be used to clean air pollution. Using static electricity to add an electric charge to dirt particles in the air can control air pollution from factory smoke. The polluted air is then passed through a chamber with positively charged electrodes, and the negatively charged dirt particles stick to the electrodes leaving the remaining air free of dirt.

What's Online

Try these websites to find out more about electrical charges and energy.

http://www.mos.org/sln/toe/staticmenu.html

http://www.mos.org/sln/toe/toe.html

CONVERTING ELECTRICAL ENERGY

What You'll Find

TERMS AND DEFINITIONS

Circuit

Definition: A circuit is a complete path for electrical charges to move from one place to another.

In Context: When the light switch is off, the circuit is broken, and the light is turned off.

Conductor

Definition: A conductor is a substance that allows electricity and heat to flow easily through it.

In Context: Electrical wire is made of metal because it is a good conductor.

Insulator

Definition: An insulator is a substance that does not allow electricity to flow through it.

In Context: The plastic around an electrical wire is an insulator and does not allow electricity to pass through it.

Your World and Science

Sitting at home with your family on a stormy night when the lights go out is scary but fun. Sometimes the power goes out when it is very windy or when there is a lot of lightning. If the storm damages an electrical line that connects to your home, the electricity in your house will turn off. If an electrical line is blown down in a storm, the path that electricity follows from the power plant to your house has been broken. Without a complete path, electrical currents cannot travel into your home, and the lights go out. That's when it's time to light the candles, cuddle on the couch, and tell scary stories. And don't forget to eat all the chocolate ice cream in the freezer because without electricity the refrigerator stops working.

What You Need to Know

In the last section, you learned that static electricity builds up in an object when negatively charged electrons move to the object. The electricity is non-moving or static until it is released onto a positively or neutrally charged object like a doorknob. When the electric charges move, they become electric current. For the electric current to keep moving, it must have a path to follow.

CIRCUITS

A circuit is a path that allows electrical energy to flow from one place to another. There are two main types of circuits, a series circuit and a parallel circuit. A series circuit provides only one path for the electricity to follow. If there is a break in the path, the current cannot flow. For example, a long string of holiday lights on a series circuit will work only if all the lights are working. If one of the lights burns out, the circuit path is broken, and none of the lights will work.

Another example of a circuit is inside a flashlight. If you stack two good batteries correctly into the handle of a flashlight, the light will turn on when the switch is moved. The batteries are electric cells that contain chemical energy to move charges through a circuit. The switch on the flashlight lets you control the flow of current by turning the flashlight on or off. When the switch is on, the path is complete. Then electricity can travel to the tiny light bulb inside the flashlight. But if the switch is off, a break occurs in the path, and the bulb will not light.

In a parallel circuit, the electric current has more than one path to follow. If a strand of lights were on a parallel circuit, the lights would stay lit even if one of the bulbs went out. There is another path for the electricity to follow, so the rest of the lights will stay lit.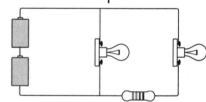

Conductors

Have you ever wondered why metal is used to make wires for electricity? Many metals have freely moving electrons floating around the nucleus of their atoms. Because of these loose electrons, metals make good conductors of electricity. A conductor allows electric current to flow easily. The wires inside the plug on your television allow electricity to flow freely into your TV set.

INSULATORS

Insulators are materials that do not allow electricity to flow easily. The wires in your house are covered with heavy plastic material. Plastic does not let electric currents flow through it. The plastic acts as an insulator so that electricity cannot flow from the wires to you! Electricity is important, and it is dangerous. Without insulators to control the flow of electric current, electricity would be too dangerous to use inside our homes.

Careful!

Humans are conductors and not insulators. This means that electricity can flow easily through your body. Never touch an exposed electrical wire or put a metal object into an electrical outlet. Keep electrical appliances away from water because water is a great conductor.

Let's Try It!

(Answer key page 206)

Einstein Questions

Can you remember what you read? Fill in the blanks. Use your highlighter to find key words in the text above.

1. A series circuit allows _____ path for the electricity to follow.

2. A conductor allows _____ to flow easily.

3. The plastic around wires acts as an _____.

Super-Genius Questions

Choose whether the material is a conductor or an insulator. Circle one.

4. Water   Conductor   Insulator

5. Plastic   Conductor   Insulator

6. Metal   Conductor   Insulator

Mental Marvel Questions

Choose the correct answer.

7. The switch on a flashlight _____.
   A. contains electricity
   B. should be made of metal
   C. turns the music louder
   D. opens and closes the circuit

8. To be safe around electricity _____.
   A. never touch an exposed wire
   B. do not take your electric radio into the bathtub
   C. never put a metal fork into the toaster
   D. all of the above

Think About It

Use the Frankenstein website listed below to discover why you should never fly a kite near power lines.

Be a Scientist

Create an electrical safety poster for your family. Provide easy to follow safety tips with colorful pictures. Make sure to include the dos and don'ts of using electricity. Post it on your refrigerator so your whole family can see it. Use the websites below to help.

Food for Thought

One streak or bolt of lightning has a huge amount of electricity. The amount of current generated by one streak of lightning is enough to provide electrical service to 200,000 homes for some time.

If you are ever diving in the ocean, watch out for electric eels. One eel can produce as much as 650 volts of electricity. Ouch!

What's Online

Try these websites to find out more about converting electrical energy.

http://www.fema.gov/kids/thunder.htm

http://www.miamisci.org/af/sln/frankenstein/index.html

Chapter 8

Objects in the Sky

THE BIG IDEA

The stars, planets, and moon move in a regular pattern in the sky.

THE SUN AND STARS

What You'll Find

TERMS AND DEFINITIONS

Axis

Definition: The **axis** is the line on which a planet or star turns.

In Context: The Sun turns on its **axis**.

Rotation

Definition: **Rotation** is an act of spinning around on an axis.

In Context: It takes 24 hours for the Earth to turn one **rotation** on its axis.

Orbit

Definition: To **orbit** is to move on a path around an object.

In Context: It takes one year for the Earth to **orbit** around the Sun.

Equator

Definition: The **equator** is the invisible circle that divides the Earth into two equal halves, north and south.

In Context: The **equator** is an equal distance between the North Pole and South Pole.

Your World and Science

Probably one of the first songs you learned to sing was "Twinkle, Twinkle Little Star." When you were younger, you may have looked up into the night sky and thought "Wow! Those stars are tiny." But now that you are older, you know that those stars are not at all tiny. In fact, many of them are so big there is almost no word to describe their size. Gigantic, colossal, and humongous just don't sound big enough for them. When you look into the night sky, you might see more stars depending on the weather, the season, and where you live. But the stars are always there in a pattern that never changes. And during the day, there is only one star that we can see. As we have learned, it is the main source of energy for all of the life on Earth. That's right. It's the Sun!

Careful!

The Sun is not a planet. It is a star. Never look straight at the Sun. It can hurt your eyes.

What You Need to Know

Scientists believe that the Sun is about 4.5 billion years old. It is by far the biggest object in our solar system. In fact, there is enough room inside the Sun to fit over a million Earths. Inside its center, the Sun is burning hydrogen. This makes the heat and light that comes up to the Sun's surface and then to our planet. The surface of the Sun is nearly 10,000 degrees Fahrenheit.

As big as the Sun is, it is still only considered a medium-sized yellow dwarf star. There are other stars that are much bigger and much hotter! The Sun is closer to Earth than any other star, so scientists can learn a lot about stars by studying our Sun. As you learned in Chapter 7, Earth has a magnetic field around it. The Sun's magnetic field is very large and very active. The Sun also has a very strong force of gravity. This gravity keeps all of the planets in our solar system in orbit around it. The Sun is the center of the solar system. Earth rotates on its axis and, at the same time, it revolves around the Sun. But because Earth's axis is tilted or tipped to one side instead of straight up and down, we get changes in weather. The tilt also makes days longer and shorter. The tilt of the Earth's axis affects the way the Sun's light hits the surface of the Earth.

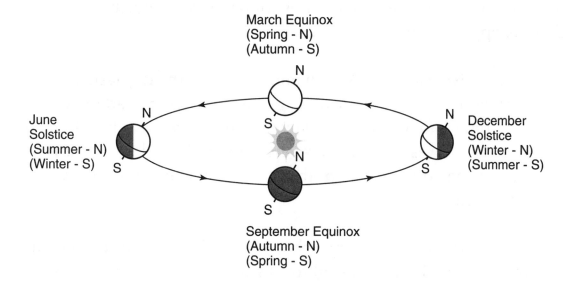

An invisible line called the equator divides Earth into two halves, north and south. When the northern half is tipped away from the Sun, it is winter in that half of the planet. And the southern half of the Earth is in summer. When the northern half is tipped toward the Sun, it is summer in the north and winter in the southern part of the world. That is why there are four seasons on our planet.

Picture this—You live in North America, which is on the northern half of the planet. It is late September, and the day is exactly as long as the night. The leaves are turning red and orange. Many leaves are falling on the ground, and you have to rake them into a pile for your dad. The air outside is getting colder so you pull on your favorite jacket. It is fall, and the northern half of the planet is beginning to tip away from the Sun.

Picture this—Three months later, you step outside in your coat and mittens. The temperature is very cold, and there is some snow on the ground. You only have a short time to play because it will be dark early. There are still 24 hours in the day, but the Sun shines for fewer hours. Even in the middle of the day, the Sun seems low in the sky, and it does not warm the air or the ground. It is winter, and the northern half of Earth is tipped away from the Sun.

Picture this—Now it is late March and, as in fall, the day lasts as long as the night. But this time the air is getting warmer instead of colder. It is almost time to put away your winter clothes. Trees are beginning to fill with green leaves and birds are returning to their old nests. It is spring, and the northern half of the planet is beginning to tip back toward the Sun.

Picture this—It is late June, and school has ended. You have big plans to swim, bike, and eat ice cream for the next three months. The Sun is high in the sky, and the air and sidewalk under your feet are hot! It is summer, and the northern half of the planet is tipped toward the Sun. Of course, this means that the southern half of the planet is in winter. Too bad for them, you think, as you kick off your shoes and run through the sprinkler.

BILLIONS OF STARS

Hundreds of years ago, before scientists used telescopes to study the sky, people counted on the stars for many things. Once people realized that the stars were always in a set pattern in the night sky, they began to name some of the brighter stars. The brightest stars were then used to predict the coming changes in seasons. This allowed the first calendars to be made. Sailors and ship captains counted on the stars to lead them in the right direction. People on land, like Native Americans, used the stars to help them move from place to place. Some ancient cultures even included the stars in their religious beliefs.

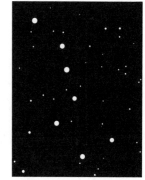

Let's Try It!

(Answer key page 207)

Einstein Questions

Can you remember what you read? Fill in the blanks. Use your highlighter to find key words in the text above.

1. The Sun has a very strong force of _____.

2. The Earth _____ around the Sun.

3. The Earth's axis is _____.

Super-Genius Questions

Decide if the statement is **true** or **false**.

4. The Sun is the largest object in the solar system. True or False

5. The Earth spins on its equator. True or False

6. When the northern half of a planet is tilted toward the Sun, it is winter in the north. True or False

Mental Marvel Questions

Choose the correct answer.

7. The Sun is considered a _____.
   A. supergiant red star
   B. yellow dwarf star
   C. supernova
   D. giant blue star

8. The change in seasons is caused by the _____.
   A. tilt of the Earth's axis
   B. the equator
   C. the orbit of the Sun
   D. the stars in the sky

9. In old days, people used stars to _____.
   A. find their way on the ocean
   B. predict changes in the season
   C. migrate from place to place on land
   D. all of the above

Think About It

What days of the year do the summer and winter solstices fall on? What days of the year do the fall and spring equinoxes fall on? If you need help go to the websites listed below to find out more.

Be a Scientist

One of the easiest ways to track the movement of the Sun is to look at shadows. Place a tall stick in a sunny place. Look at the shadow the stick is making on the ground early in the morning. Check it again at noon and once as the Sun is beginning to set in the late afternoon. Record your observations in your journal. When was the shadow the longest? When was it the shortest? Did the time of day affect the place where the stick's shadow fell on the ground? Why do you think the shadow changed as time passed through the day?

Food for Thought

Even though the Sun is massive in size compared to the planets, the gravitational pull from the planets causes the Sun to wobble a tiny bit on its axis. This wobble allows scientists to know whether or not a star has planets orbiting around it.

Life Cycle of a Star

Most stars like our Sun began life as a cloud of gas and dust. Gravity pulls the atoms of gas and dust together, and the temperature in the center grows hotter. If the star is big enough, it begins to shine brightly. It becomes a yellow or red dwarf star like our Sun. This stage of a star's life lasts about 5 billion years. When the star is old and near the end of its life, it grows much bigger, and the temperature cools down. Now the star is a red giant or even a supergiant star. Most red giant stars then become white dwarf stars until they die out completely. The supergiant stars have short lives, and at the end, some explode! This is called a supernova. It is a very powerful explosion in space. When these supergiants die, they leave a strong force of gravity behind. Scientists believe this is how the black holes in space are made.

What's Online

Try these websites to find out more about the Sun and other stars.

http://www.kidsastronomy.com/our_sun.htm

http://www.kidscosmos.org/kid-stuff/sun-facts.html

http://liftoff.msfc.nasa.gov/News/2001/News-AutumnalEquinox.asp

THE MOON

What You'll Find

TERMS AND DEFINITIONS

Satellite

Definition: A satellite is a natural body that orbits around a planet.

In Context: The moon is the Earth's only natural satellite.

Waxing

Definition: Waxing is a slow growth in size.

In Context: After the new moon, the waxing moon grows larger as it travels around the Earth.

Waning

Definition: Waning is a slow shrinking in size or strength.

In Context: After the full moon, the waning moon shrinks as it travels around the Earth.

Your World and Science

It's a clear, bright night. A full moon is shining down on your backyard campout. You and your friends are huddled together telling ghost stories, when suddenly you hear a howl in the distance. Was it a coyote or perhaps a werewolf coming out to hunt by the light of the full moon? You decide there is no such thing as werewolves but still think it's time to get ready for bed in your safe tent. Wait! What was that? A bat flying across the moon or was it a vampire? You convince yourselves that it is just your imagination or maybe too much chocolate milk. The full moon has nothing to do with werewolves or vampires. In fact, your mother suggested having your campout on a night with a full moon so it would not be so dark. What was that? Rustling in the bushes next door? Okay, there is no such thing as monsters, but just in case, you and your friends decide to move the campout to your room.

What You Need to Know

There are many myths and legends connected to the moon. People all over the world have celebrations connected to the phases of the moon. Since the beginning of time, people have been fascinated with the moon. It is the biggest, brightest object in the night sky. But let's cut through all the stories and legends surrounding the moon and get to the facts.

Moon Facts

- The moon is the Earth's only natural satellite.

- The moon is about 240,000 miles away from the Earth, making it our closest neighbor.

- The moon takes about 29 days or about a month to make one orbit around the Earth.

- It also takes the moon about one month to rotate or spin on its axis.

- The moon has no atmosphere and has a gravitational force that is 1/17th that of the Earth's.

- The moon is not made of cheese, and there is no man in the moon.

PHASES OF THE MOON

Although the moon is always traveling around the Earth, it can't always be seen. If you were to watch the moon every night for a whole month, the moon would look as if it were growing larger and then shrinking again. The Sun is always shining on half of the moon. The changes in the moon's size and shape depend on how much of the lighted surface of the moon can be seen from the Earth. These changes in the way the moon looks are called the phases of the moon. The moon travels around the Earth every 29½ days, and during that time it goes through eight phases.

New Moon

During this phase, the moon cannot be seen. This is because the Sun lights the side of the moon that does not face the Earth. The only time the moon can be seen during this phase is if there is a solar eclipse. A solar eclipse is when the Earth, moon, and Sun are perfectly lined up so that the moon blocks out the Sun. At that time, the moon can be seen as it passes in front of the Sun.

Waxing Crescent

As the moon moves around the Earth, we begin to see the lit surface. During this phase, only a very small piece of the moon's edge can be seen from Earth. This crescent or small piece of the moon is curved.

First Quarter

During this phase, half of the moon appears to be lit by the Sun.

Waxing Gibbous

The moon is still growing during this phase. It now looks almost full.

Full Moon

The side of the moon facing the Earth is now completely lit by the Sun.

Waning Gibbous

At this point in the moon's trip around the Earth, the moon looks as if it's shrinking. It looks slightly less than full.

Last Quarter

During this phase, half of the moon looks lit by the Sun.

Waning Crescent

During this phase, only a very small part of the moon can be seen. The moon continues to shrink as it completes its trip around the Earth.

The next phase after the waning crescent moon is the new moon, and the phases begin again. It takes one month for the moon to go through all eight phases. Each change happens slowly and may look as if it lasts a couple of days.

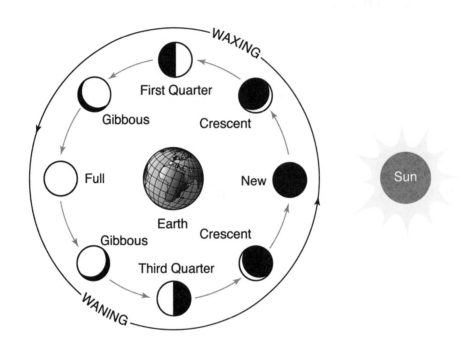

Let's Try It!

(Answer key page 207)

Einstein Questions

Can you remember what you read? Fill in the blanks. Use your highlighter to find key words in the text above.

1. The moon is Earth's only _____.

2. It takes the moon_____ to orbit the Earth.

3. The moon is invisible during the _____ phase.

Super-Genius Questions

Match the word with its definition.

4. Full moon
5. First quarter
6. Waning gibbous
7. Waxing crescent

A. The moon is shrinking. It looks almost full.
B. The moon is completely lit.
C. The moon is growing. A very small piece can be seen.
D. Half the moon appears to be lit.

Mental Marvel Questions

Choose the correct answer.

8. The moon shines brightly in the night sky because _____.

   A. it is made of yellow cheese
   B. it is a star
   C. it reflects the Sun's light
   D. it is made of glow in the dark material

9. During the waxing gibbous phase, the moon is _____.

   A. shrinking in size
   B. full
   C. shaped like a crescent
   D. growing in size

Think About It

Write a short description of the surface of the moon. If you need help, go to the websites listed below to find out more.

Be a Scientist

Now it's your turn to take a closer look at the moon. You will need your science journal, a pencil, clear skies, and a pair of binoculars if possible. Once you have your materials and have picked a good night for moon watching, you may begin your observation. Draw a circle in your journal and record the moon phase you see by shading in the part of the moon you do not see. If you have binoculars, now would be a good time to get a closer look to add detail to your drawing. Make sure to list the time and date you made your observation. Wait about a week and pick another night to observe the moon. Repeat the steps you took during the first observation. Continue the observation for the entire 4-week cycle.

Food for Thought

A billion years ago, the moon was much closer to the Earth. Its orbit around the Earth only took 20 days. Over time, the moon orbit has grown larger, and it has slowed down.

The same side of the moon always faces the Earth.

What's Online

Try these websites to find out more about the moon.

http://www.enchantedlearning.com/subjects/astronomy/ moon/Phases.shtml

http://www.enchantedlearning.com/subjects/astronomy/ moon/

http://www.promotega.org/ksu30002/moon_act.htm

http://stardate.org/nightsky/moon/

THE PLANETS

What You'll Find

TERMS AND DEFINITIONS

Planet

Definition: A **planet** is a heavenly body traveling in an orbit around the Sun.

In Context: Earth is a **planet** that travels around the Sun.

Asteroid

Definition: An **asteroid** is any of thousands of smaller objects that travel around the Sun.

In Context: Most **asteroids** orbit the Sun between Mars and Jupiter.

Craters

Definition: A **crater** is a bowl-shaped dent made by a meteor in the surface of a planet.

In Context: Earth's moon is covered in **craters**.

Atmosphere

Definition: The **atmosphere** is the layer of gases that surround a planet.

In Context: The Earth's **atmosphere** has many gases including oxygen.

Your World and Science

What? Pluto's no longer a plant! What next? Chocolate's no longer a flavor? As you sit in your classroom, listening to your teacher rip away the one sure thing you knew about science, you wonder if anything she says can be trusted. Have you not been taught since kindergarten that the solar system has nine planets and one Sun? Now, suddenly, it has eight planets. Has the number of Suns changed too? Just like that, a group of adults can get together and kick Pluto out of the planet club. If they're going to get together and change anything, couldn't it be that broccoli is no longer good for you?

What You Need to Know

The solar system consists of one Sun and eight planets. Yes, that's eight planets. As of August 2006, the International Astronomical Union (the people in charge of all the information on space and the solar system) decided that Pluto is no longer a planet. As we are able to see and reach farther out into the solar system, what we know about the solar system changes. That is why Pluto has been renamed as a dwarf or minor planet.

THE INNER PLANETS

Picture this—You are going to take a tour of the solar system, but your mom wants you home by dinner. Traveling by spaceship, it would take you 3 months to get to Venus, the closest planet to Earth and 12 years to get to Neptune, the furthest. Mom won't go for that. So you decide to use a transporter to travel at the speed of light to get there faster.

The solar system is divided into inner planets and outer planets separated by an asteroid belt. You are going to visit the four inner planets first. They are all warmer than the outer planets because they are closer to the Sun. They are all small planets with rocky surfaces. You set the controls on your transporter for Mercury, the closest plant to the Sun.

Mercury is about the size of Earth's moon, but it has no moon of its own. Mercury has almost no atmosphere to protect it, so its surface is marked by large craters made by asteroids hitting it. You would love to land, but the temperature on the surface of the planet facing the Sun is hot enough to melt metal. You could land on the side facing away from the Sun, but the temperature would be extremely cold, and you forgot your sweater.

Venus is the next planet on your list. It is the second closest planet to the Sun. This planet is about the same size as Earth and is sometimes called Earth's twin. But there are many differences between the two planets. Venus has no moon. It is also covered with craters and volcanoes. You could not land here even if you wanted to. Your ship would melt either from the heat or from the acid that is part of the atmosphere. Venus is not a friendly place.

The next inner planet is Earth. This is a planet you know well. It is the right distance from the Sun, with the right atmosphere and water to support life. And since it is home, you can explore it any time. So you go on to the next planet, Mars.

Mars is the fourth planet from the Sun. As you fly down through the Martian atmosphere, you notice that everything looks reddish orange. This is because the soil on Mars is full of iron. It is as if the surface is covered in rust just like your bike when you leave it out in the rain. As you look down, you see the solar system's tallest volcano and deepest valley. You look up into the Martian sky and wonder if there ever were real Martians to enjoy the view of Mars's two moons.

The Outer Planets

To get to the outer planets, you have to go through the asteroid belt. This is an area between Mars and Jupiter that is about 620 miles wide. It contains pieces of early planets that did not form when the solar system began over 4.6 billion years ago. The four outer planets have many things in common. They are all really cold gas giants. They all have rings of ice and dust and many moons.

Past the asteroid belt, you come to the fifth and largest planet in the solar system, Jupiter. Jupiter is a giant gas planet made up of hydrogen and helium gas. The atmosphere on Jupiter is very stormy. The largest storm has lasted for hundreds of years and looks like a big red spot. In fact, it is called the Great Red Spot. Flying past Jupiter, you can see the thin rings that circle the planet as well as the moons that orbit it. Jupiter has as many as 63 moons surrounding it. Wow! That is a lot of moons.

The next planet you come to as you continue your trip away from the Sun is the sixth planet, Saturn. It is known as the ringed planet. Saturn is also a giant gas planet like Jupiter, and its atmosphere is stormy too. As you get closer to the planet, you notice that the planet looks like it is squashed on the top and bottom and bulging at the middle. Saturn spins so quickly on its axis that it looks as if it is flattened out. Saturn is a beautiful sight. It has about 1000 rings. The rings may look solid from space, but they are really made up of thousands of broken pieces of moons and asteroids. The pieces come in a variety of sizes from as tiny as a pebble to as large as a house.

Since you promised your mom you'd be home for dinner, you had better get to the last two planets in the solar system. First is Uranus, planet number 7. The first thing you notice is that it looks as if it is tipped over on its side. Uranus is really cold. The cold methane gas that is part of its atmosphere gives the planet its beautiful blue color. This planet also has rings and many moons.

Flying by planet number 8, Neptune is a lot like looking at Uranus all over again. They are very much alike. They are both giant gas planets with rings and many moons. But Neptune has a large, dark spot that scientists believe is a storm much like Jupiter's. One interesting thing about Neptune is that Triton, one of its moons, orbits the planet in the opposite direction of its other moons.

Before you head home, you decide to fly past Pluto so you don't hurt its feelings. Even though Pluto is no longer a planet, it is still an object in the solar system. Pluto is so cold that its atmosphere lays frozen on its surface. Scientists have been arguing whether Pluto is a planet since it was discovered. Many scientists believe that Pluto is one of Neptune's moons that escaped its gravitational pull. Recently, it has been discovered to be part of the Kuiper Belt, a group of icy objects orbiting the Sun at the farthest end of the solar system.

Now it is time to go home. As much as you have enjoyed visiting the other planets in the solar system, there is no place like Earth.

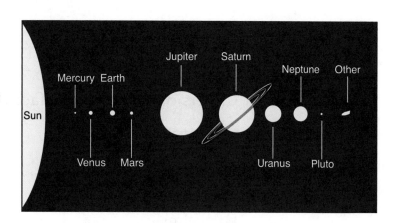

Let's Try It!

(Answer key page 207)

Einstein Questions

Can you remember what you read? Fill in the blanks. Use your highlighter to find key words in the text above.

1. The closest planet to the Sun is _____.

2. Mars is the _____ planet from the Sun.

3. There is an _____ between Mars and Jupiter.

Super-Genius Questions

Place the planet under the correct heading.

	Inner planet	Outer planet
4. Venus		
5. Neptune		
6. Earth		
7. Jupiter		

Mental Marvel Questions

Choose the correct answer.

8. Craters on a planet's surface are caused by

_____.

   A. really big footsteps
   B. meteors or asteroids
   C. earthquakes
   D. oceans

9. The Great Red Spot on Jupiter's surface is caused by
a _____.
A. giant storm
B. really big paint spill
C. volcano
D. moon

Think About It

Why do you think there is life on Earth but not on any other planet in our solar system? If you need help, go to the websites listed below to find out more.

Be a Scientist

Copy the chart below into your journal. Use the websites below to fill in the information about each planet.

Name of Planet	Miles from the Sun	Length of Day (rotation period)	Length of Year (revolution period)
Mercury			
Venus			
Earth			
Mars			
Jupiter			
Saturn			
Uranus			
Neptune			

Food for Thought

All the planets in the solar system rotate counterclockwise, except Venus. Venus rotates clockwise.

Scientists have found at least 34 Martian meteorites on Earth. They have been studying these meteors for evidence of ancient bacteria or other forms of life that could have existed on Mars.

What's Online

Try these websites to find out more about the planets.

http://www.exploratorium.edu/ronh/age/

http://starchild.gsfc.nasa.gov/docs/StarChild/solar_system_level1/planets.html

Chapter 9

Rocks and Minerals

THE BIG IDEA,

Rocks and minerals are formed in the Earth over time.

CLASSIFYING ROCKS

What You'll Find

TERMS AND DEFINITIONS

Pressure

Definition: **Pressure** is the force on the surface of an object.

In Context: The boy used his hand to put **pressure** on a bleeding cut.

Magma

Definition: **Magma** is hot, melted rock within the Earth.

In Context: There is a river of **magma** that flows under the Earth's surface.

Sediment

Definition: **Sediment** is material like sand and rocks that have been moved by water or wind.

In Context: The bottom of the river was full of **sediment**.

Igneous

Definition: **Igneous** rock is formed when lava or magma cools to a solid.

In Context: **Igneous** rock can be found in and around a volcano.

Metamorphic

Definition: **Metamorphic** change is a change in form caused by pressure and heat.

In Context: Heat can change igneous rock into **metamorphic** rock.

Your World and Science

You're on a field trip to look for fossils, but all you have found are some boring rocks. Boring! Rocks aren't boring. Think about it. Rocks come in all shapes and sizes from tiny grains of sand to huge boulders. Rocks come in many colors, even striped and speckled. Rocks can tell scientists a lot about the Earth and the animals and plants that lived on it long ago. Rocks aren't boring. They have a lot to say. You just have to learn to listen.

What You Need to Know

Even though rocks come in many shapes, sizes, and colors, there are only three kinds of rocks. Igneous, sedimentary, and metamorphic are the three kinds of rock. Each of these three types of rock is made in a different part of the rock cycle. The rock cycle is a group of changes. To understand the rock cycle better, you need to take a trip through it.

Igneous Rock

You begin your trip deep in the Earth, in the middle of a volcano. This is where igneous rock begins as magma. Magma is hot, melted rock that sits under the surface of the Earth. Igneous or fire rock, as it is sometimes called, can form above the ground or below the ground. Igneous rock forms underground when magma becomes trapped in pockets and starts to cool very slowly. Granite is a type of igneous rock that forms underground. Igneous rock can also form above ground. When a volcano erupts, magma rises to the surface of the Earth. Magma above ground is called lava. As lava pours from a volcano, it will cool quickly, forming igneous rock. Pumice is a type of igneous rock that forms from lava above ground.

Careful!

Magma and lava are the same thing, but the name changes from magma to lava when magma flows above the ground.

Sedimentary Rock

Once the igneous rock is formed and sitting on the surface of the Earth, wind and water begin to break it down into smaller pieces. The little pieces of rock that are broken down are called sediment. The sediment gets washed down to the bottom of rivers, lakes, and oceans. Over thousands of years, layers of sediment pile up causing pressure and weight on the bottom of the pile. This causes the sediment on the bottom of the pile to press together creating sedimentary rock. Sandstone and limestone are both types of sedimentary rock that are made by small grains of different minerals pressed together in layers.

METAMORPHIC ROCK

The next and last stop on your trip around the rock cycle is once again deep in the Earth. But this time you are not going to be near magma. Instead, one kind of rock will change into a different kind of rock, just like a caterpillar into a butterfly. Metamorphic rock is rock that changes. Rocks change when they are under heat and pressure. Rocks buried deep in the Earth are under pressure from tons of other rock piled on top of them. The pressure and the heat inside the Earth cause rocks to change. The heat comes from the Earth's plates shifting and coming together. You will learn more about Earth's plates in the next chapter. The coming together causes friction just as rubbing your hands together will make heat. The mineral crystals that make up the rock may change and flatten out making a new rock. Schist is metamorphic rock that comes from shale, a sedimentary rock. Gneiss rock is metamorphic rock that may once have been granite, an igneous rock.

To complete the trip around the rock cycle, metamorphic rock can be melted back into magma to eventually became igneous rock or be broken down into sediment to become sedimentary rock.

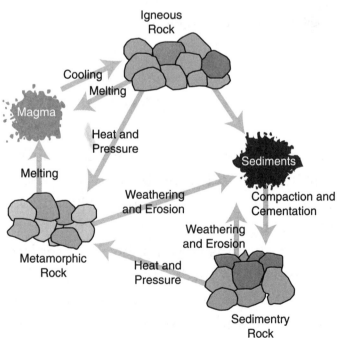

Let's Try It!

(Answer key page 208)

Einstein Questions

Can you remember what you read? Fill in the blanks. Use your highlighter to find key words in the text above.

1. Igneous rocks are made _____ or _____ the ground.

2. Little pieces of broken-down rock are called _____.

3. Metamorphic rock is rock that _____.

Super-Genius Questions

Match the name of the rock with its type.

4. Gneiss   A. igneous

5. Granite   B. sedimentary

6. Limestone   C. metamorphic

Mental Marvel Questions

Choose the correct answer.

7. Igneous rock starts out as _____.
   A. sediment
   B. other rock
   C. magma
   D. water

8. Sediment is made by _____.
   A. water and wind
   B. water and magma
   C. lava and clouds
   D. all of the above

Think About It

What is the difference between the way igneous rocks and metamorphic rocks are made?

Be a Scientist

Now it's your turn to be a rock hound. The best way to learn about rocks is to go out and find some. Go out to your backyard, school playground, or park. Find rocks that are interesting to you. Then fill in the chart. You can use the websites in this book or other library books on rocks to help identify your samples.

	Sample 1	Sample 2	Sample 3	Sample 4
Draw the rock you found				
Where did you find it?				
What shape is it?				
What colors do you see?				
How does it feel? (rough or smooth)				
Is it an igneous, sedimentary, or metamorphic rock?				

Food for Thought

If you wanted to dig a hole through to the opposite side of the Earth, it would take you 87 years of digging at a rate of one foot a minute.

The oldest known rock on Earth can be found in Canada. It is a 3.96 billion year old Acasta gneiss rock.

Many of the rocks found on Earth come from space. In fact, as many as 100,000 tons of rocks fall to Earth from space each year.

What's Online

Try these websites to find out more about classifying rocks.

http://www.minsocam.org/MSA/K12/rkcycle/ rkcycleindex.html

http://www.cotf.edu/ete/modules/msese/earthsysflr/ rock.html

http://www.childrensmuseum.org/geomysteries/ mysteries.html

http://library.thinkquest.org/J002289/index.html

MINERALS

What You'll Find

TERMS AND DEFINITIONS

Classify

Definition: To **classify** is to arrange or organize in like groups.

In Context: Scientists **classify** minerals according to their color and hardness.

Crystal

Definition: When a solid substance has a repeating pattern of atoms, it is a **crystal**.

In Context: We saw a beautiful piece of quartz **crystal** at the rock show.

Streak

Definition: A **streak** is a line of powder left behind when a mineral is rubbed on a hard, white surface.

In Context: Some minerals are classified according to the color of the **streak** or mark they leave on a white surface.

Physical Property

Definition: A **physical property** is anything you can observe about an object.

In Context: The hardness of a mineral is a **physical property**.

Your World and Science

What do the salt on your French fries, the sand on the beach, and the diamond in your mom's engagement ring all have in common? They are all minerals. Minerals are all around you. They are in the ground you walk on, the food you eat, and the water you drink.

What You Need to Know

Scientists have discovered and named about 3000 minerals on Earth. Minerals are not plants or animals. They are solids that grow in the Earth. Minerals get pressed together to form rocks and mountains. Calcite is one of the most common minerals on Earth. It is found in limestone and sedimentary rocks. Minerals like iron and calcium are used in medicine. Most foods contain minerals like salt and potassium. Animals and people need these minerals to live. Building materials like steel and iron all come from minerals. Gemstones used in jewelry are made from minerals like diamond and ruby. Under the right conditions, minerals grow into repeating patterns and forms called crystals. They are all different, and they can be identified by their physical properties. Physical properties like color and hardness are different for every mineral. Scientists use these physical properties to classify all the minerals on Earth.

Testing Physical Properties

Color

Minerals come in many different colors. Some minerals always have the same color. Gold is an example of a mineral that always has the same color. Some minerals, like quartz and calcite, have different colors depending on what's inside them. Minerals with copper inside will look green. Iron will make minerals look purple or dark brown. Sometimes different mineral crystals grow on top of each other making different colors appear. So color is not always the best physical property for scientists to use when classifying minerals.

Streak

A better way to classify a mineral is to do a streak test. When a mineral is rubbed across the smooth white surface of a tile, it leaves behind a streak of colored powder. The streak may be a different color than the mineral itself. Scientists can classify the mineral by the color of the powder streak left behind.

Hardness

Have you ever heard that a diamond is hard enough to scratch glass? That's because diamond is the hardest mineral on Earth. When a diamond is not pretty enough for jewelry, it is used to top off the ends of drills. Drills are tools that are used to dig through hard surfaces like rock. Diamonds are that hard. But some minerals are not hard. Scientists can classify minerals by their hardness. Some minerals are soft enough to be scratched by a fingernail. Some minerals cannot be scratched by a fingernail, but they can be scratched with the side of a coin. Some minerals are too hard for the side of a coin, but a knife or a piece of glass can scratch them. A hammer can destroy diamonds and quartz, but they are too hard to scratch with a knife. The hardest mineral, diamond, is given a hardness score of ten. Talc, one of the softest minerals, is given a hardness score of one.

Let's Try It!

(Answer key page 208)

Einstein Questions

Can you remember what you read? Fill in the blanks. Use your highlighter to find key words in the text above.

1. Scientists use physical properties to _____ all the minerals.

2. _____ are the hardest mineral found on Earth.

3. Minerals grow in repeating patterns and geometric forms called _____.

Super-Genius Questions

Decide if the statement is true or false.

4. A diamond can be scratched with a fingernail. True or False

5. Minerals always have the same color. True or False

6. Minerals can be plants growing on the Earth. True or False

Mental Marvel Questions

Choose the correct answer.

7. A mineral that contains the element copper might look _____.
   A. scary
   B. yellow
   C. clear
   D. green

8. Besides jewelry, diamonds are sometimes used to make _____.
   A. tools
   B. cakes
   C. other minerals
   D. stars

9. The softest minerals can be scratched with a _____.
   A. feather
   B. fingernail
   C. whisper
   D. all of the above

Think About It

Some scientists use luster as a way to classify minerals. What is luster? If you need help, go to the websites listed below to find out more.

Be a Scientist

Rock candy is the crystal form of sugar. To make rock candy, you will need 2 cups of sugar, 1 cup of water, a small pot, a wooden spoon, a clean glass jar, a piece of string, a metal screw, a craft stick, and your journal. And the most important ingredient is an adult to help you because you will be cooking with heat. Put 1 cup of water and 2 cups of sugar into the pot and place it on the stove. Heat the mixture until the water comes to a boil and the sugar is completely dissolved. Remember to keep stirring while the solution is heating. Have an adult pour the solution into the clean jar. Dip a piece of string in the sugar solution and then pull it back out. Let the sugar-soaked string dry over night. The next day, tie one end of the string around the craft stick and the other end of the string around the metal screw. The screw will act as a weight to keep the string down in the sugar water. The craft stick should be placed across the top of the jar so that the tip of the string stays out of the solution. Place the jar in a quiet place where it will not be moved or tipped over. After several days, you should see some sugar crystals begin to form on the string. In your journal record your findings each day for a week. Draw a picture of the crystals as they begin to form rock candy on the string. After crystals have formed, they can be removed from the solution and eaten. If your parents say you can. And if you do eat it, watch your teeth. Sugar crystals are hard.

Food for Thought

Mineral crystals do not just form on rocks. Our bones are made up of tiny crystals of a mineral called apatite.

Why are perfect diamonds so rare and valuable? First of all, diamond crystals must have perfect conditions to grow. After they grow, they must remain perfect. But water, rocks, and other forces in nature make it hard for diamonds to stay protected. As they slowly move to Earth's surface, they must stay safe from miners who dig the diamonds out of caves. Miners use hammers and dynamite to blast through rocks. Many diamonds are destroyed in the mining process. The few that survive are the lucky ones, and they are worth a great deal of money.

What's Online

Try these websites to find out more about minerals.

http://mwoolley.customer.netspace.net.au/top.htm

http://library.thinkquest.org/J002744/adlm.html

http://www.loelem.santacruz.k12.ca.us/classrooms/ room16/projectsfieldtrips/facts.html

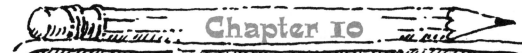

Chapter 10

Shaping Earth's Surface

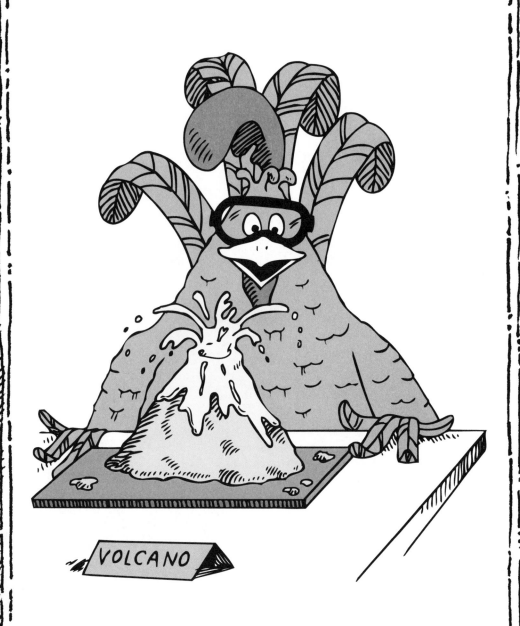

VOLCANO

THE BIG IDEA

Earth's surface is constantly changing.

EARTHQUAKES AND VOLCANOES

What You'll Find

TERMS AND DEFINITIONS

Plates

Definition: **Plates** are large pieces of Earth's crust that are constantly moving.

In Context: The Earth's crust is made up of many large **plates**.

Fault

Definition: A **fault** is a break or crack in a large body of rock.

In Context: Earthquakes may happen along **faults** in the Earth's surface.

Colliding

Definition: **Colliding** objects are objects hitting each other with strong force.

In Context: Two cars were **colliding** on the freeway in a terrible accident.

Your World and Science

You are sitting on the beach building a cool sandcastle with your sister. Your creepy cousin is sitting on a beach chair eating the entire bag of potato chips by himself. Even though you built your dream castle away from the water, some of the salty ocean washes up on shore and takes away an edge of your wall. No big deal. Now it looks even cooler. Your creepy cousin has finished his chips and is licking the salt off his chubby fingers. Suddenly he lets out an evil laugh. You drop your shovel and bucket and watch in horror as he races across the sand and leaps onto your sandcastle. The whole thing is ruined. A major disaster has hit your castle, and now it is nothing but a flat pile of wet sand. The Earth can sometimes act the same way as your cousin. Most of the time, Earth's changes are gentle like the water that washed away the edge of your castle wall. But sometimes Earth can be dangerous and destructive like your creepy cousin. And when Earth feels like being destructive—watch out!

What You Need to Know

The Earth is made up of three layers. Deep in the center is a solid core of iron. Above that is a thick layer of magma or melted rock. And the part we walk on is the crust. Earth's crust is very thin like the brown crust on a loaf of bread. The crust is not one solid piece. It is a giant puzzle of big pieces or plates that fit together around the entire planet. Since these pieces of crust are sitting on liquid magma they are always moving. Think of what it might be like to build a puzzle on a giant bowl of Jell-O. Although the pieces are always moving, we don't usually feel the movement because the pieces are big, and the movement is slow.

Sometimes the plates push into each other or pull away from each other. Sometimes the plates rub against each other. When this happens, they change the surface of the Earth. Sometimes mountains are formed. Sometimes canyons appear, and sometimes a new volcano is formed. The movement of Earth's plates can be much more destructive than your creepy cousin.

EARTHQUAKES

Picture this—For this imaginary trip through time you will become a fly. Why, you might ask. The place you are going is about to be hit by a big earthquake, and you are much safer as a flying insect. The year is 1964, and you are a fly on the wall of a department store in Alaska. You are wearing your tiny fly ski coat because the weather is very cold. It is Alaska after all. You are about to dive for a piece of hot dog lying on the ground when you notice that the floor is rolling like waves on the ocean. The walls begin to shake, and there is a loud rumble moving through the town. People are screaming and running into the street. The building jumps sideways, and part of the wall crumbles into the street.

After several long minutes of rocking and rolling, the ground has stopped moving. People come out of their hiding places shaking and crying. You decide to fly out the open side of the building to see what has happened. The street is no longer one flat road. It has split apart in many places, and cars have fallen into the deep ditches left behind. Huge rivers of mud and snow are flowing down the surrounding hillsides into the town below. Water pipes have broken open, and the streets are being flooded. You have just witnessed the 1964 Alaskan earthquake. It was one of the biggest ever recorded in North America. Many buildings were damaged, and people were hurt or killed. Some of the worst damage and danger came from the giant ocean waves created by the quake.

How Did It Happen?

We have learned that the Earth's crust is made up of giant pieces or plates that are always moving. Plates that make up the land of the seven continents are called the continental plates. Plates that make up the floors of Earth's oceans are called oceanic plates. There are also deep cracks in the Earth's crust called faults. These cracks can be anywhere on the plate where there is a break in the solid rock.

When two plates slide against each other from side to side, they release energy. The energy released causes the Earth's crust to move. This movement causes an earthquake. Sometimes one plate will slip under another plate also causing an earthquake. This is what happened on the coast of Alaska in 1964. Sometimes energy is released along a fault, and this will cause the Earth to shake too. Once the energy is released between the two plates or along a fault, the movement rolls out under the crust in waves. Think about what happens when you drop a rock into a pond of water. Ripples move out from the place where the rock hit the water. Earthquake waves move in the same way.

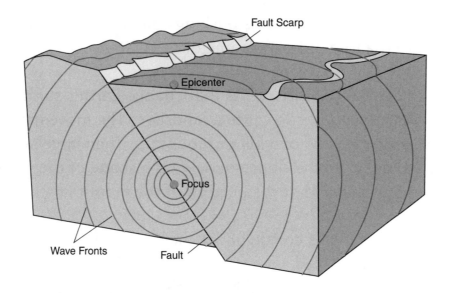

There are at least a million earthquakes around the world every year. But most of them are too small to cause damage. The size of the earthquake depends on how much energy is released by the two colliding plates. Sometimes plates touch each other and the earthquake is too small for people and animals to feel. Sometimes the plates collide deep down on the ocean floor, far away from land. When two plates move against each other and release a lot of energy, the earthquake can be very big and dangerous. People who live near the place where the two plates struck each other could be in danger of losing their homes. When an earthquake is strong enough, buildings can fall, and the ground can break open.

CHANGES IN THE EARTH'S SURFACE

When two big plates crash together, the crust of the Earth is pushed up to form mountains. The Himalayas in Asia are the highest mountains on Earth. They formed when two continental plates collided. Some mountains, like the Grand Tetons in the United States, formed in the middle of a plate. The Grand Tetons were pushed up in the middle of a plate by the pressure of one plate pushing against another. The pressure caused a fold or bulge in one of the plates.

Earthquakes can cause other events that change Earth's surface. If the ground moves enough, it can cause a landslide. The movement can shake loose an entire side of a hill or mountain. Then large amounts of dirt or mud slide down the slope changing the shape of the mountain and the shape of the land at the bottom of the mountain.

When two oceanic plates collide under the sea and the ocean floor shakes, a giant wave called a tsunami or tidal wave may form in the ocean. If the movement was close enough to land, the tidal wave could wash onto land and cause a lot of damage and even death. Many times the giant waves cause more damage than the earthquake.

Sometimes plates in the Earth's crust pull apart instead of pushing together. When this happens, a new volcano is formed.

Volcanoes

Picture this—It is 79 A.D. and you are a fly on the stone wall, in the middle of an open theater, in the ancient Roman city of Pompeii. You are watching the busy town trying to decide what tasty piece of food to land on next when a giant explosion cracks through the air. The theater walls shake wildly. Behind you, the mountain known as Vesuvius looks like it is on fire. Huge clouds of smoke and flames shoot out from the top of the mountain. You watch as people come out of their homes and shops to stare up at the exploding mountain. Some people are packing their wooden carts with food and supplies as if they plan to leave the town. You decide that it would be best to follow the food. A town without food is no place for a fly.

You zoom down to a cart packed with fresh bread and find a safe place to land. The streets are now filled with large crowds of people. The streets are so packed that people are moving slowly. Everyone is pointing back to the mountain. There is a large, dark cloud moving toward the town. Lightning strikes begin flashing through the center of the cloud. Suddenly, the people begin pointing to the ocean. The sea looks as if it has been sucked backwards away from the land. There are thousands of fish gasping for breath on the dry land. Now the cloud of smoke and dust begins to cover the entire town. It is hard to breathe, and the people begin to move faster through the crowded streets. It is the middle of day, yet it is as dark as the darkest night. The people are moving too slowly, and you fly up through the air and away from the giant cloud of smoke and ash. There is sure to be some better food in the next town, and you have had enough of volcanoes.

How Did It Happen?

We have learned that earthquakes happen when two of Earth's plates push together. But sometimes plates move away from each other. This makes a big gap in the Earth's crust. And as we have learned, beneath the crust lies a giant layer of hot magma. As the Earth's crust breaks open, the hot magma begins to flow up to the surface as lava and ash. As it cools, the lava and ash become solid mountains. This is how a volcano is formed.

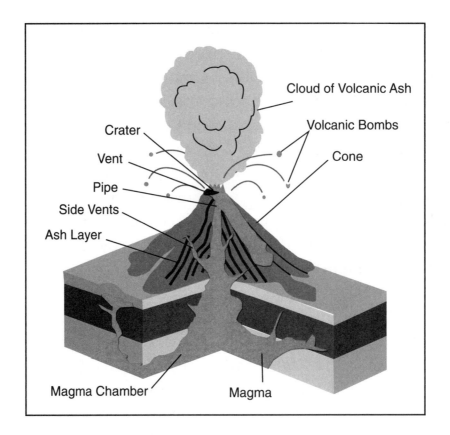

Cloud of Volcanic Ash

Volcanic Bombs

Crater

Cone

Vent

Pipe

Side Vents

Ash Layer

Magma Chamber

Magma

Around the Pacific Ocean there are long chains of volcanoes that have formed when the oceanic plate of the Pacific Ocean slipped under the continental plates. Some volcanoes, like the ones that make up the Hawaiian Islands, have formed in the middle of a plate. The plate is always floating over a thick layer of hot magma. When a hot column of magma pushes through the plate, it causes a volcanic eruption. The hot lava pours out on the surface of the plate, and a new volcano is formed.

Let's Try It!
(Answer key page 209)

Einstein Questions

Can you remember what you read? Fill in the blanks. Use your highlighter to find key words in the text above.

1. An _____ happens when two plates in Earth's crust collide.

2. When two plates in the Earth's crust pull apart, a _____ is formed.

3. A column of hot magma pushing through a plate causes a volcanic _____.

Super-Genius Questions

Decide whether the statement is **true** or **false**.

4. Mountain ranges are formed when a tidal wave hits land. True or False

5. A volcano can form when two of Earth's plates pull apart. True or False

6. A tidal wave can form after a big earthquake. True or False

Mental Marvel Questions

Choose the correct answer.

7. When two plates collide, they release _____.

    A. lava
    B. energy
    C. crust
    D. water

8. The plates on Earth are always _____.
   A. moving
   B. jumping
   C. standing very still
   D. all of the above

Think About It

What is the largest volcano in the world? If you need help, go to the websites listed below to find out more.

Be a Scientist

Now it is time to do some research. You will need a map of the world and access to a library and the Internet. Find the locations of some of the world's volcanoes and draw them on your map. You could color active volcanoes red and extinct volcanoes black. Don't forget to label them.

Food for Thought

The biggest earthquake that scientists were able to record happened in 1960 in Chile, South America. Scientists estimated the earthquake to be a 9.5 on the magnitude scale. Fortunately, not many lives were lost because there had been several small foreshocks before the big one hit and people had already left their buildings and homes.

There are as many as 1500 active volcanoes on land in the world today, but there are thousands more under the ocean.

What's Online

Try these websites to find out more about earthquakes and volcanoes.

http://www.learner.org/exhibits/volcanoes/

http://volcano.und.nodak.edu/

http://www.sfusd.k12.ca.us/schwww/sch618/Volcanoes/Volcanoes.html

http://www.weatherwizkids.com/earthquake1.htm

EROSION

What You'll Find

TERMS AND DEFINITIONS

Weathering

Definition: Weathering is a process of changing rock by wind and water.

In Context: Weathering has changed the surface of the Earth.

Deposition

Definition: **Deposition** is the act of dropping or laying something in a new place.

In Context: The Mississippi Delta was made by the **deposition** of sediment.

Erosion

Definition: **Erosion** is a process of wearing away the surface of the Earth.

In Context: **Erosion** makes changes to mountains and beaches everyday.

Your World and Science

Your family is spending the day in the mountains. You will have a picnic, then hike the trails to look for interesting rocks and leaves. After you eat, you hike through a cool forest kicking at the leaves that cover the ground. The rocks and dirt crunch under your feet. You pick up a rock and decide to break it open to see what it looks like on the inside. You use it like a hammer and hit it against a big rock. After a few hard hits, it finally breaks open. With all the whacking, the rock does not just break in half. Both rocks have a lot of little chunks broken off of them. You are now part of the erosion process, breaking big rocks into little rocks.

What You Need to Know

The surface of the Earth is always changing. It changes all the time and every day in big and small ways. In the last section, you read about how the Earth's surface can change suddenly with the eruption of volcanoes or the shaking of earthquakes. But in this section, you will learn about the slow, small changes that happen with erosion. Erosion is a process of breaking down rock and land into smaller and smaller pieces. Mountains are broken down everyday by water, wind, plants, and animals. As the rocks are chipped away from mountains, they are pushed downhill by gravity and running water from streams and melting snow. As you learned in Chapter 9, rocks and land washed down stream are called sediment, and sediment does not just disappear. It gets built up into another form.

WEATHERING

Picture this—You are lying in your bed at night and suddenly, you hear strange noises. It sounds like someone is in the house, or maybe it's a ghost moaning and groaning. Relax. It is only your house settling. It was a really warm day, and the wood in the walls of your house expanded or grew bigger with the heat. As the night cools, the walls contract or shrink back down. That's what is making the sound. Well guess what? Rocks do the same thing. They can expand and contract with temperature changes. But as rocks grow and shrink in the heat and cold, they start to crack and break down.

Water is an important part of erosion. Fast moving rivers can wear away rock and soil, making deep canyons in the land. Water, in the form of ocean waves, can wash away sand from beaches. Waves also wear away rocks from cliffs, changing the shape of the coast.

Water can also make its way into tiny cracks in rocks. If the temperature gets cold enough, the water will freeze. Water expands as it turns into ice causing the cracks in the rocks to grow wider. As the temperature grows warmer, the ice melts, leaving a bigger crack, and the process begins again. More water will run into the cracks and freeze. Over time this causes the rocks to break apart. This is the freeze-thaw cycle.

Plants and animals can also cause erosion. Plants that grow in the soil have roots that grow into tiny cracks in rocks. Remember when you learned about simple machines and wedges? These roots act as wedges as they grow bigger, breaking rocks apart over time. Animals help the erosion and weathering process by digging in the ground to make their dens and burrows.

TRANSPORT AND DEPOSITION

What happens to all the rock and earth that has been broken down by weathering? Water, wind, and gravity move or transport the rock and earth to a new place. Rivers, rain, and melting snow carry rock and mud down mountains to the ocean where it is dropped or deposited. Over many years, the sediment builds up into new land such as wetlands and deltas. Ocean waves not only erode cliffs with their pounding action but also deposit the pieces of rock as new sand on the beach.

Wind helps with the transport and deposit of sediment. Wind blows dry, loose sediment into new land forms. Wind coming off the ocean can blow sand into dunes along the beach. These dunes protect the coastline during a storm.

Erosion is the way Earth recycles itself. Mountains are broken down, and the sediment moves to a different place and forms a new type of land. It is not something that happens quickly like an earthquake. It is a slow, steady process that takes years.

Let's Try It!

(Answer key page 209)

Einstein Questions

Can you remember what you read? Fill in the blanks. Use your highlighter to find key words in the text above.

1. Erosion is caused by _____, _____, _____, and _____.

2. Temperature can cause rocks to _____ and _____.

3. _____ can get into tiny cracks in rocks and freeze.

Super-Genius Questions

Decide if the statement is **true** or **false**

4. Temperature changes can cause a rock to crack. True or False

5. Wind can move wet, heavy rocks to new places. True or False

6. Plants do not cause erosion. True or False

Mental Marvel Questions

Choose the correct answer.

7. Rock that has been eroded is transported and deposited by _____.
   A. gravity, plants, and animals
   B. wind, animals, and gravity
   C. lava, plants, and water
   D. water, wind, and gravity

8. Animals help with erosion by _____.
   A. hunting
   B. digging
   C. climbing trees
   D. eating grass

Think About It

There are other types of weathering besides the ones discussed above. Name one. If you need help, go to the websites listed below to find out more.

Be a Scientist

Now it's your turn to experiment with erosion.

Question: *How does rain help to change the surface of the Earth?*

Hypothesis: Remember a hypothesis is your guess or prediction about what will happen in the experiment. Make sure to read through the experiment to understand it before you write your hypothesis in your journal.

Materials: You will need 2 pans, soil, 2 or 3 books, a watering can, and your journal.

Procedures: Stack the books one on top of the other. Place one pan on the table next to the stack of books. Prop one end of one of the pans up on the books. Put the other end into the second pan. This will put the pan at a slant like the side of a hill. Then fill the pan that is propped up with soil to the rim. Pack the soil down tightly. Carefully pour water from the can on to the top end of the pan and let it run down.

Record: What has happened to the soil? Record exactly what happened in your journal.

Conclusion: In your conclusion, you should answer the question: "How does rain help to change the surface of the Earth?" You also need to explain why your hypothesis was right or wrong.

Food for Thought

Cape Hatteras lighthouse in North Carolina has been moved because of erosion. The erosion did not come from the ocean waves but from the tourists who come every year and spit over the rail of the observation deck.

The water going over Niagara Falls has eroded 7 miles from the bottom of the falls over the last 10,000 years.

Scientist have estimated that it took between 6 million and 8 million years for erosion to make the Grand Canyon in Arizona.

What's Online

Try these websites to find out more about erosion.

http://www.geography4kids.com/files/land_weathering.html

Answer Key

CHAPTER 1

Page 5

Einstein Level
1. protection
2. cracking open nuts
3. webbed

Super-Genius Level
4. C   6. A
5. B

Mental Marvel Level
7. B
8. C

Think About It
Possible answers—Walruses have large tusks to help them climb out of the water onto the ice. They also use their tusks to anchor themselves to the bottom of the ocean while they dig for clams.

Walruses can squirt high-powered jets of water to help them uncover clams on the ocean floor.

Walruses have sacs under their throats that they fill with air to help them float in water to sleep. They also have blubber to help keep themselves warm in the cold Pacific Ocean.

Page 13

Einstein level
1. B   3. A
2. D   4. C

Super-Genius Level
5. C
6. B

Mental Marvel Level
7. PF   10. PF
8. PF   11. SB
9. SB

Think About It
Possible answer—Foxes use their claws to dig dens in the forest.

Page 20

Einstein Level
1. run faster
2. crocodiles, sharks
3. extinct

Super-Genius Level
4. False   6. False
5. True

Mental Marvel Level
7. C
8. D

Think About It
Possible answer—Horses' legs grew longer so they could run from predators because there were fewer bushes to hide in.

CHAPTER 2

Page 28

Einstein Level
1. Chlorophyll
2. sugars
3. holes

Super-Genius Level
4. D   6. B
5. A   7. C

Mental Marvel Level
8. B
9. D

Think About It
Possible answer—No, because mushrooms do not have chlorophyll.

Page 38

Einstein Level
1. Plants
2. 3, 4
3. plants, animals

Super-Genius Level
4. B
5. C

Mental Marvel Level
6. cactus→lizard→hawk
7. flower→insect→salmon→bear
8. plankton→fish→seal→killer whale

Think About It
Possible answer—Sharks, whales, and dolphins are all top-level carnivores in the ocean.

Page 44

Einstein Level
1. decomposers
2. mushrooms, molds
3. smallest

Super-Genius Level
4. False   6. True
5. False

Mental Marvel Level
7. B
8. D

Think About It
Possible answer—Yes, earthworms eat dead plants and animals.

CHAPTER 3

Page 54

Einstein Level
1. plants, animals
2. decomposers
3. pollution

Super-Genius Level

    4. mouse

    5. river

    6. pine tree

Mental Marvel Level

    7. living

    8. non-living

    9. living

    10. non-living

Think About It

Possible answers—Three living things are toucans, bats, and butterflies. Three non-living things are soil, rain, and rocks.

Page 59

Einstein Level

    1. Dispersal

    2. food, shelter

    3. pollination, seed dispersal

Super-Genius Level

    4. False   6. False

    5. True

Mental Marvel Level

    7. A

    8. D

Think About It

Possible answer—Besides seed dispersal by animals, seeds are moved by wind, by moving water, and by the plants themselves.

CHAPTER 4

Page 70

Einstein Level

    1. energy

    2. Sun

    3. food, fossil fuels

Super-Genius Level

    4. True   6. True

    5. False

Mental Marvel Level

    7. A

    8. B

Think About It

Possible answer—Renewable energy does not cause pollution or waste.

Be a Scientist

Conclusion—The black bottle is hotter and the balloon will blow up on the black bottle. It should be concluded that the warm air inside the black bottle floated up into the balloon because the hot air rises.

Page 77

Einstein Level

    1. waves

    2. water, sound, light

    3. Transfer

Super-Genius Level

4. True   6. True
5. False

Mental Marvel Level

7. B
8. D

Think About It
Possible answer—Light waves when the light turns on, the light reflecting off the mirror, the water moving in the bathtub, the sound of your voice in the bathroom.

Page 84

Einstein Level

1. fulcrum
2. wedge
3. groove

Super-Genius Level

4. A   6. B
5. D   7. C

Mental Marvel Level

8. B
9. D

Think About It
Possible answer—The can opener (lever), the string for opening the blinds (pulley), the ladder (incline plane), and the wagon (wheels and axles).

Be a Scientist
Conclusion—Friction does affect an object moving on an inclined plane. The sandpaper increases the friction so it takes more energy to push the paper clip up and down the inclined plane.

CHAPTER 5

Page 95

Einstein Level

1. Matter
2. space
3. Molecules

Super-Genius Level

4. C   6. B
5. A

Mental Marvel Level

7. C
8. D

Think About It
Possible answers—Plasma and Bose-Einstein condensates are two other forms of matter.

Page 102

Einstein Level

1. gas
2. heat, cold
3. freezing point

Super-Genius Level

    4. A   6. B

    5. C

Mental Marvel Level

    7. C

    8. A

Think About It

Possible answer—The freezing point of water is 0 degrees Celsius or 32 degrees Fahrenheit. The boiling point of water is 100 degrees Celsius or 212 degrees Fahrenheit.

CHAPTER 6

Page 110

Einstein Level

    1. straight

    2. transparent

    3. shadow

Super-Genius Level

    4. False   6. True

    5. False

Mental Marvel Level

    7. B

    8. A

Think About It

Possible answer—The reflection of sunlight in water drops in the air causes a rainbow.

Page 114

Einstein Level

    1. reflected

    2. Refraction

    3. speed

Super-Genius Level

    4. True   6. True

    5. False

Mental Marvel Level

    7. C

    8. A

Think About It

Possible answer—During reflection, light bounces off an object. During refraction, light goes through an object and bends.

Be a Scientist

Possible answer—Light bends when it moves from air to water because the water molecules are more tightly bound than the air molecules. Light slows down when it enters the water.

CHAPTER 7

Page 124

Einstein Level

    1. attract

    2. magnetic fields

    3. battery

Super-Genius Level
    4. False   6. True
    5. True

Mental Marvel Level
    7. A
    8. D

Think About It
Possible answer—The school bell, car, blender, TV, and radio.

Be a Scientist
Possible answer—The iron nail as a core would make the strongest magnet because electrons are packed loosely in metal and metal is a good conductor. Plastic does not allow electrons to flow through it and does not make a good core in an electromagnet.

Page 131

Einstein Level
    1. protons, neutrons
    2. negative
    3. equal

Super-Genius Level
    4. B   6. C
    5. A

Mental Marvel Level
    7. C
    8. A

Think About It
Possible answer—The shock on your brother's head, the hair sticking to your head after putting on your sweater, the comb causing your hair to stick up, and the underwear sticking to your sweater.

Be a Scientist
Possible answer—The balloons eventually fall off when the rubbed off electrons have moved back to the balloons, making the charge neutral.

Page 137

Einstein Level
    1. one
    2. electricity
    3. insulator

Super-Genius Level
    4. Conductor
    5. Insulator
    6. Conductor

Mental Marvel Level
    7. D
    8. D

Think About It
Possible answer—The kite could fly into the power lines, and electricity could travel down the string on the kite and shock you.

CHAPTER 8

Page 147

Einstein Level
1. gravity   3. tilted
2. orbits

Super-Genius Level
4. True   6. False
5. False

Mental Marvel Level
7. B   9. D
8. A

Think About It
Possible answer—The summer solstice is on or around June 21. The winter solstice is on or around December 21. The spring equinox is on or around March 21. The fall equinox is on or around September 21.

Page 154

Einstein Level
1. satellite
2. 29½ days
3. new moon

Super-Genius Level
4. B   6. A
5. D   7. C

Mental Marvel Level
8. C
9. D

Think About It
Possible answer—The surface of the moon has craters and lots of rocks. No water can be seen.

Page 162

Einstein Level
1. Mercury
2. fourth
3. asteroid belt

Super-Genius Level
4. Inner planet
5. Outer planet
6. Inner planet
7. Outer planet

Mental Marvel Level
8. B
9. A

Think About It
Possible answer—Earth is just the right distance from the Sun. It is not too hot or too cold. It has a better atmosphere and it has water.

Be a Scientist

Name of Planet	Miles from the Sun	Length of Day (rotation period)	Length of Year (revolution period)
Mercury	57.8 million km	58.6 days	87.9 days
Venus	108.2 million km	243 days	224.7 days
Earth	152 million km	23.93 hours	365.26 days
Mars	228 million km	24.6 hours	686.98 days
Jupiter	778.3 million km	9 hours 55 minutes	11.78 years
Saturn	1.429 billion km	10 hours 40 minutes	29.46 years
Uranus	2.871 billion km	17 hours 14 minutes	84.01 years
Neptune	4.501 billion km	16.11 hours	164.79 years

CHAPTER 9

Page 171

Einstein Level
1. above, below
2. sediment
3. changes

Super-Genius Level
4. C   6. B
5. A

Mental Marvel Level
7. C
8. A

Think About It
Possible answer—Igneous rock is made from lava or magma. Changing other rocks with heat and pressure makes metamorphic rock.

Page 177

Einstein Level
1. classify
2. Diamonds
3. crystals

Super-Genius Level
4. False   6. False
5. False

Mental Marvel Level
7. D   9. B
8. A

Think About It
Possible answer—Luster refers to the way a mineral reflects light. Some minerals may be non-metallic, and some may be metallic.

CHAPTER 10

Page 191

Einstein Level
1. earthquake
2. volcano
3. eruption

Super-Genius Level
4. False   6. True
5. True

Mental Marvel Level
7. B
8. A

Think About It
Possible answer—The Mauna Loa volcano in Hawaii is the world's largest.

Be a Scientist
Possible answer—Mt. Ranier and Mt. St. Helens are located on the continent of North America. Mt. Vesuvius and Mt. Aetna are located on the continent of Europe. The Fogo Caldera volcano is located on the continent of Africa. The Pinatubo volcano is located on the continent of Asia.

Page 197

Einstein Level
1. water, wind, plants, animals
2. expand, contract
3. Water

Super-Genius Level
4. True   6. False
5. False

Mental Marvel Level
7. D
8. B

Think About It
Possible answer—Salt wedging can occur when salt water seeps into rocks, salt crystals form, and the rock cracks apart.

Be a Scientist
Possible answer—The soil should have been pushed down the hill into the second pan. Rain helps erosion by washing away soil.

INDEX